SpringerBriefs in Materials

W0230338

More information about this series at http://www.springer.com/series/10111

The SpringerBriefs Series in Materials presents highly relevant, concise monographs on a wide range of topics covering fundamental advances and new applications in the field. Areas of interest include topical information on innovative, structural and functional materials and composites as well as fundamental principles, physical properties, materials theory and design. SpringerBriefs present succinct summaries of cutting-edge research and practical applications across a wide spectrum of fields. Featuring compact volumes of 50 to 125 pages, the series covers a range of content from professional to academic. Typical topics might include:

- A timely report of state-of-the-art analytical techniques
- A bridge between new research results, as published in journal articles, and a contextual literature review
- A snapshot of a hot or emerging topic
- An in-depth case study or clinical example
- A presentation of core concepts that students must understand in order to make independent contributions

Briefs are characterized by fast, global electronic dissemination, standard publishing contracts, standardized manuscript preparation and formatting guidelines, and expedited production schedules.

Amit Chauhan

Non-Circular Journal Bearings

 Springer

Amit Chauhan
Panjab University
Chandigarh
India

.

ISSN 2192-1091   ISSN 2192-1105   (electronic)
SpringerBriefs in Materials
ISBN 978-3-319-27331-0   ISBN 978-3-319-27333-4   (eBook)
DOI 10.1007/978-3-319-27333-4

Library of Congress Control Number: 2015957399

Springer Cham Heidelberg New York Dordrecht London

Printed on acid-free paper

Springer International Publishing AG Switzerland is part of Springer Science+Business Media (www.springer.com)

Contents

About the Author

Amit Chauhan is an assistant professor at the Mechanical Engineering Department at UIET, Panjab University, Chandigarh, India. He graduated with a Ph.D. in mechanical engineering from the National Institute of Technology, Himachal Pradesh, India, in 2011. He has published one book on vertical axis wind turbine and has authored or coauthored over 50 published articles. His research interests are in the area of tribology with particular interest in noncircular journal bearings, composite materials, and nonrenewable energy sources. He is a life member of the Tribology Society of India and member of the High Energy Materials Society of India.

Chapter 1
Introduction

Hydrodynamic bearings are known as the mechanical components that smoothly support the external loads due to geometry and relative motion of mating surfaces when a thick film of lubricant exists between them. Such bearings find extensive use in high-speed rotating machines as they have low friction, high load capacity, and good damping characteristics. Due to high speed of the rotor, there was the tendency of occurrence of oil whip and oil whirl phenomenon. Formation of oil whip is not desirable because the rotor cannot form a stable wedge and it leads to metal to metal contact between rotor and bearing. Once surface contact exists, the rotor begins to precess, in a reverse direction from the actual rotor rotation direction, using the entire bearing clearance. This condition leads to high friction levels which increase the temperature of lubricating oil. This will overheat the bearing metal, thus causing rapid destruction of the bearing, rotor journal, and machine seals. Fuller (1956) has suggested that the fluid film bearings are probably the most important mechanical components in the recent technological development and are comparable in their significance (as they were used in turbines which ultimately leads to generation of power) to the effect of electricity. The development of fluid film lubrication mechanisms has been observed by Petrov (1883) in Russia and by Tower (1883) in England. In 1886, Reynolds presented his classical analysis of bearing hydrodynamics, which forms the basis of present days' bearing study. It has been seen in literature that the temperature rise in the oil film is quite high in circular journal bearings since they operate with single active oil film. To overcome the problem of excessive temperature rise, it leads to the development of bearings with non-circular profiles, which operate with more than one active oil film. The feature of operating with more than one active oil film accounts for the superior stiffness, damping, and reduced temperature in the oil film as compared to the circular journal bearings. Almost all the non-circular journal bearing geometries enhance the shaft stability under proper operating conditions. Such non-circular profile bearings will also help in reduced power losses and increase oil flow as compared to an inscribed

© The Author(s) - SpringerBriefs 2016

A. Chauhan, *Non-Circular Journal Bearings*, SpringerBriefs in Materials,

DOI 10.1007/978-3-319-27333-4_1

circular bearing, thus reducing the oil film temperature. Among non-circular journal bearings, the commonly reported in the literature are: offset-halves, two-lobe, elliptical, lemon bore, and three-lobe configurations.

References

Fuller DD. Theory and practice of lubrication for engineers. New York: Wiley; 1956.
Petrov NP. Friction in machines and the effect of lubrication. Inzh Zh St Peterburgo. 1883;1: 71–140.
Tower B. First report on friction experiments. P Inst Mech Eng. 1883;34:632–59.

Chapter 2
Classification of Non-circular Journal Bearings

The chapter presents the reasons towards the development of non-circular journal-bearing profiles and classification of such profiles. The chapter also presents the basic mechanism of operation on which non-circular journal bearing works. In the last section of the chapter, the different regime of lubrication which may occur in various types of bearing has also been presented.

2.1 Circular Journal Bearing

The basic configuration of the circular journal bearing consists of a journal that rotates relative to the bearing which is also known as bush (Fig. 2.1). Efficient operation of such bearing requires the presence of a lubricant in the clearance space between the journal and the bush. In hydrodynamic lubrication, it is assumed that the fluid does not slip at the interface with the bearing and journal surface. There exists a velocity gradient over the thickness of the fluid, which depends upon the relative movement of bearing surfaces. There will be no pressure generation if the bearing surfaces are parallel or concentric, which means bearing could support any bearing load. However, if the surfaces are at a slight angle, the resulting lubrication fluid velocity gradients will be such that generation of pressure results from the wedging action of the bearing surfaces and on the same concept hydrodynamic lubrication depends.

The operation of hydrodynamic lubrication in journal bearings has been illustrated in Fig. 2.2. Before the rotation commences, i.e. at rest, the shaft rests on the bearing surface. When the journal starts to rotate, it will climb the bearing surface gradually as the speed is further increased; it will then force the lubricant into the wedge-shaped region. When more and more lubricant is forced into a wedge-shaped clearance space, the shaft moves up the bore until an equilibrium condition is reached, and now, the shaft is supported on a wedge of lubricant. The moving

© The Author(s) - SpringerBriefs 2016
A. Chauhan, *Non-Circular Journal Bearings*, SpringerBriefs in Materials,
DOI 10.1007/978-3-319-27333-4_2

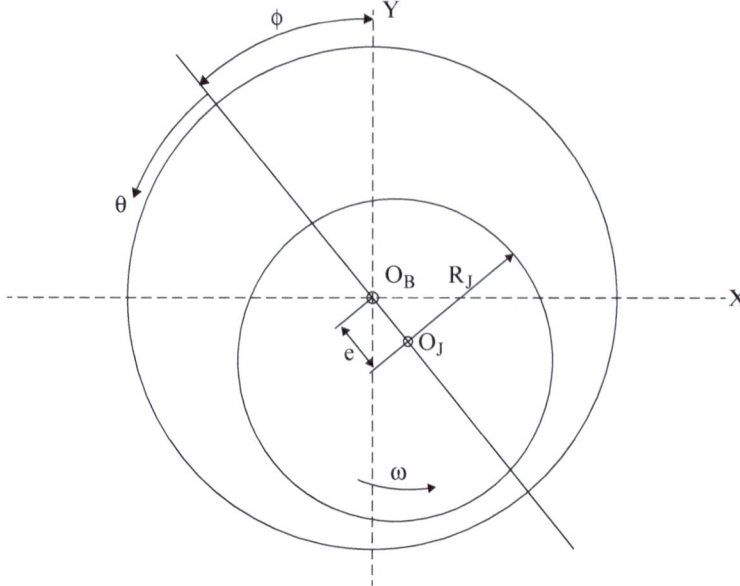

Fig. 2.1 Schematic of circular journal bearing

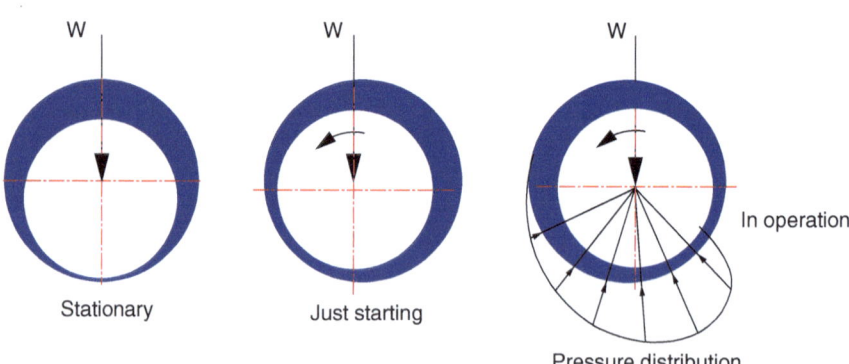

Fig. 2.2 Schematic of operation of hydrodynamic lubrication in journal bearing

surfaces are then held apart by the pressure generated within the fluid film. Journal bearings are designed such that, at normal operating conditions, the continuously generated fluid pressure supports the load with no contact between the bearing surfaces. This operating condition is known as thick film or fluid film lubrication and results in a very low operating friction.

   On the other hand, if the lubricant film is insufficient between the relatively moving parts, it may lead to surface contact and the phenomenon is normally known as boundary lubrication. This occurs at rotation start-up, a slow speed operation or if the load is too heavy. This regime results in bearing wear and a relatively high friction value. If a bearing is to be operated under boundary lubricating conditions, special lubricants must be used. Among hydrodynamic bearings, circular journal bearing is the most familiar and widely used bearing. Simple form of this bearing offers many advantages in its manufacturing as well as in its performance. However, the circular journal bearings operating at high speed encounter instability problems of whirl and whip. Instability may damage not only the bearings, but also the complete machine.

Moreover, these bearings usually experience a considerable variation in temperature due to viscous heat dissipation. This significantly affects the bearing performance as lubricant viscosity is a strong function of temperature. Furthermore, excessive rise in temperature can cause oxidation of the lubricant and, consequently, lead to failure of the bearing. Pressure also influences the viscosity of the lubricant to certain extent. Usually viscosity increases exponentially as the pressure increases, which in turn increases the load capacity of the journal bearing. Researchers have studied the behaviour of circular journal bearing by adopting various numerical approaches to simulate the performance in accordance with the real conditions.

2.2 Non-circular Journal Bearing

It has been reported in the literature that the temperature rise is quite high in circular journal bearings as they operate with single active oil film. This resulted in the development of bearings with non-circular profile, which operate with more than one active oil film. This feature accounts for the superior stiffness, damping, and reduced temperature in the oil film as compared to the circular journal bearings. Almost all the non-circular journal bearing geometries enhance the shaft stability, and under proper conditions, this will also reduce power losses and increase oil flow (as compared to an inscribed circular bearing), thus reducing the oil film temperature. Among non-circular journal bearings, offset-halves, elliptical, lemon bore, and three-lobe configurations are the most common.

The offset-halves journal bearing has been commonly used as a lobed bearing in which two lobes are obtained by orthogonally displacing the two halves of a cylindrical bearing. Offset-halves journal bearings (Fig. 2.3) find applications in gear boxes connecting turbine and generator for the power generation industries. They can also be used where primary directions of force, constant direction of rotation are found or high bearing load capacity, long service life, high stiffness, and damping values are the main features under concentration. If the unit is operated at full power, such requirements can also be met by lemon bore bearings. Lemon bore bearing is a variation of the plain bearing where bearing clearance is reduced in one direction and this bearing has a lower load-carrying capacity than the plain bearings, but is more susceptible to oil whirl at high speeds (Chauhan 2011). However, equipment must often be operated at lower performance levels, particularly in the times of reduced current needs. It is precisely under these conditions that lemon bore bearings may provide unstable conditions, which may require equipment shutdown to

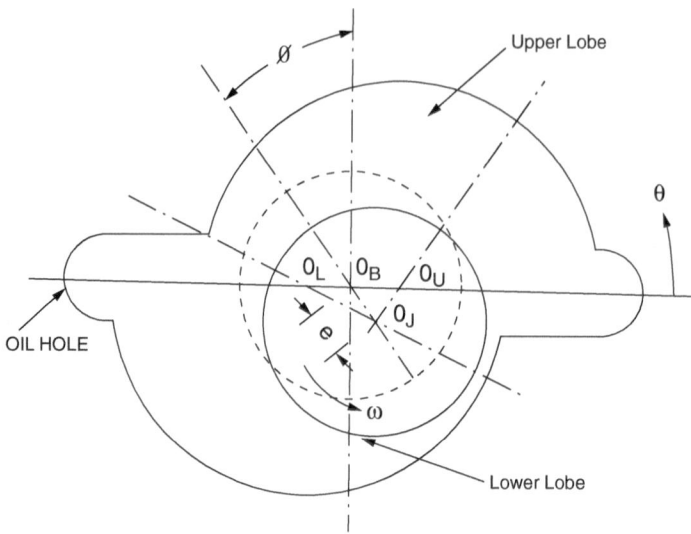

Fig. 2.3 Schematic diagram of offset-halves journal bearing

avoid damage. Offset-halves journal bearings have the durability equal to lemon bore bearings, while the bearings have good stiffness and damping properties which enables light loads while bearing runs at high speeds. The same bearing profile also offers the advantage of a long, minimally convergent inlet gap, which results in high load capacity for it. Also, the externally applied force and compression resulting from the horizontal displacement of bearing halves accurately holds the shaft in the lubricant film. This effect produces excellent characteristics such as elastic rigidity and damping by the oil film. Thus, the offset-halves journal bearings prove to be technical alternative to conventional lemon bore bearings (Chauhan and Sehgal 2008).

The elliptical journal bearings (Fig. 2.4) are commonly used in turbo-sets of small and medium ratings, steam turbines, and generators. The so-called elliptical journal bearing is actually not elliptic in cross-section but is usually made up of two circular arcs whose centers are displaced along a common vertical straight line from the centre of the bearing. The same bearing has also been referred as two-lobe journal bearing by some researchers. The bearing so produced has a large clearance in the horizontal or split direction and a smaller clearance in the vertical direction. Elliptical journal bearings are slightly more stable toward the oil whip than the cylindrical bearings. In addition to this, elliptical journal bearing runs cooler than a cylindrical bearing because of the larger horizontal clearance for the same vertical clearance. Some other non-circular journal bearing configurations; Three-lobe journal bearing (symmetrical and asymmetrical, Fig. 2.5), Elliptical journal bearing (profile is elliptic in cross-section, Fig. 2.6), and orthogonally displaced journal bearing (vertical offset, Fig. 2.7); have also been shown below.

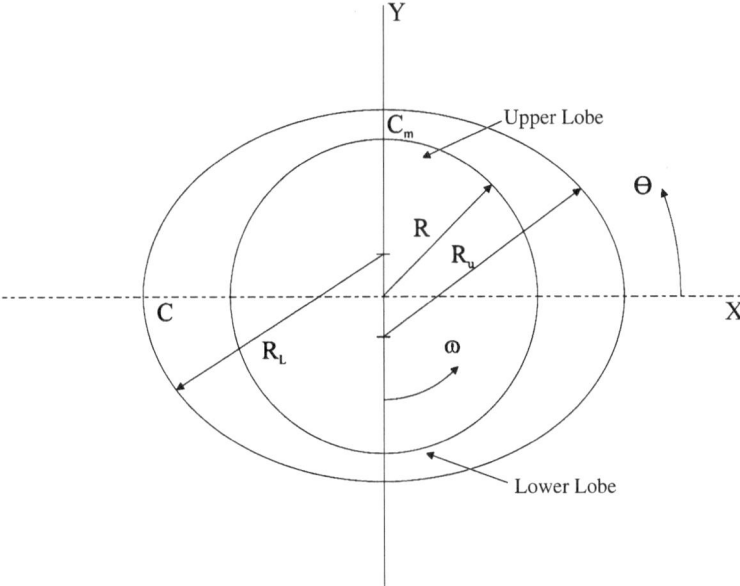

Fig. 2.4   Schematic diagram of elliptical journal bearing (two-lobe)

Fig. 2.5   Symmetrical
three-lobe bearing

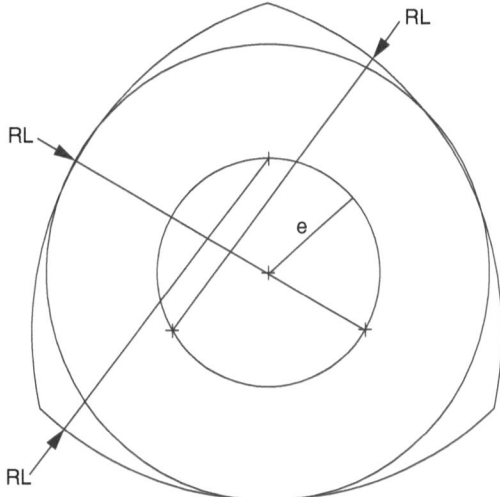

2.3   Methods of Analysis

In earlier works, the bearing performance parameters have been computed by solving the Reynolds equation only. Over the years, many researchers have proposed number of mathematical models. A more realistic thermohydrodynamic

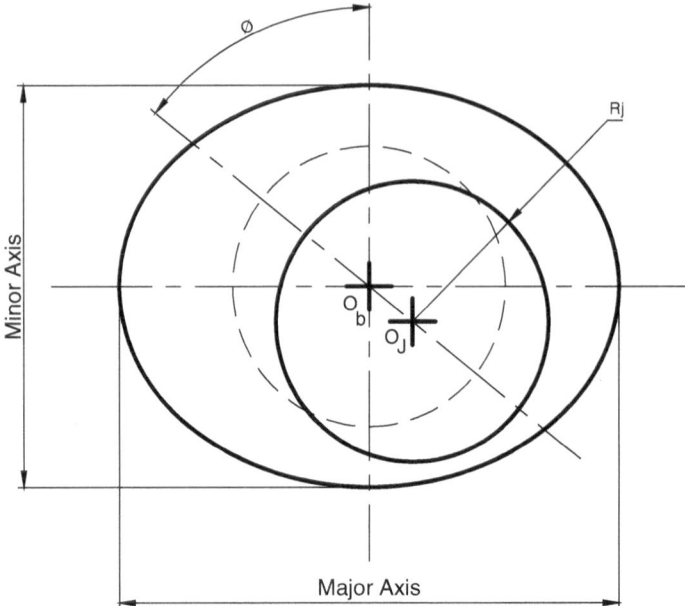

Fig. 2.6 Truly elliptical journal bearing

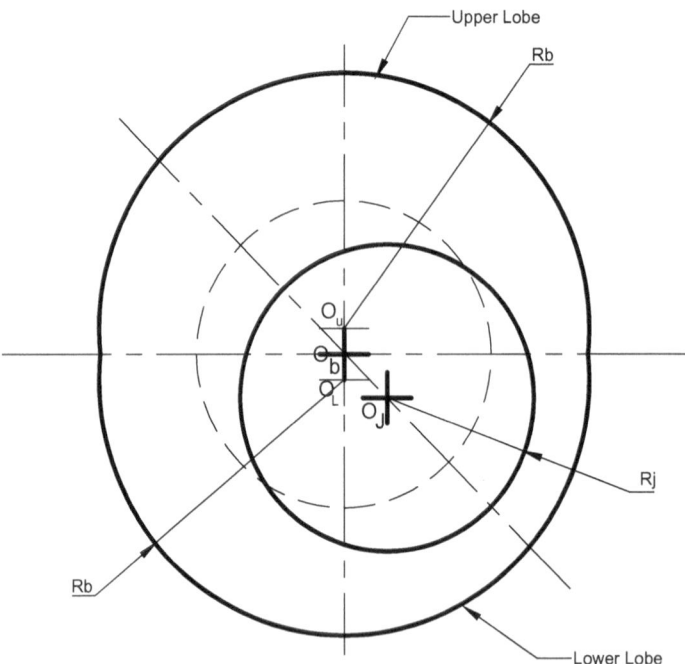

Fig. 2.7 Orthogonally displaced journal bearing

(*THD*) model for bearing analysis has been developed which treats the viscosity as a function of both the temperature and pressure. Moreover, it also considers the variation of temperature across the film thickness and through the bounding solids (housing and Journal). The thermohydrodynamic model also presents coupled solutions of governing equations by incorporating appropriate boundary conditions and considering the heat conduction across the bearing surfaces. Even the importance of *THD* studies in hydrodynamic bearings can be justified by looking at the large volumes of research papers that are being published by researchers using various models.

The theoretical investigations have been carried out into the performance of hydrodynamic journal bearing by adopting various methods, which are classified in two categories as (Kumar 2007):

1. Methods which comprise a full numerical treatment of temperature variation across the lubrication film thickness in energy equation using Finite Difference Method (FDM) or Finite Element Method (FEM).
2. Methods which incorporate polynomial approximation to evaluate the transverse temperature variation in the lubrication film thickness.

Both approaches mentioned can be used for the analysis of hydrodynamic bearings and have certain merits. The first approach is relatively accurate at the expense of computational speed and time, whereas the second is relatively fast at the expense of accuracy.

However, in recent years, the analysis works for the journal bearings have been carried out using various available simulation softwares like COMSOL, ANSYS, etc. The simulation softwares carry out the analysis in real conditions and thus the results obtained are littler higher than those obtained by above-mentioned two methods of analysis.

2.4 Regimes of Lubrication

From the ages, it is a known fact that lubrication reduces the friction between sliding surfaces by filling the surface cavities and making the surfaces smoother and such action of lubricant is known as lubrication. In other words, lubrication is a process by which the friction and wear rates in a moving contact are reduced by using suitable lubricant. Lubricant is a substance introduced between relatively moving parts to reduce friction ($\mu = 0.1$–0.0001) and wear rate. Most lubricants are some type of fluid such as mineral and synthetic oils etc.; however, there are some solid lubricants, e.g. gold, silver, polymers, etc. Almost every relatively moving component in an assembly requires lubricant. Liquid lubricants can be brought into a converging contact due to rotation and pressure generation between the bodies; they can lower the temperature of interacting surfaces and remove contaminants. Liquid lubricants can be mixed with other chemicals to provide additional properties (i.e. corrosion resistance, surface active layers, etc.).

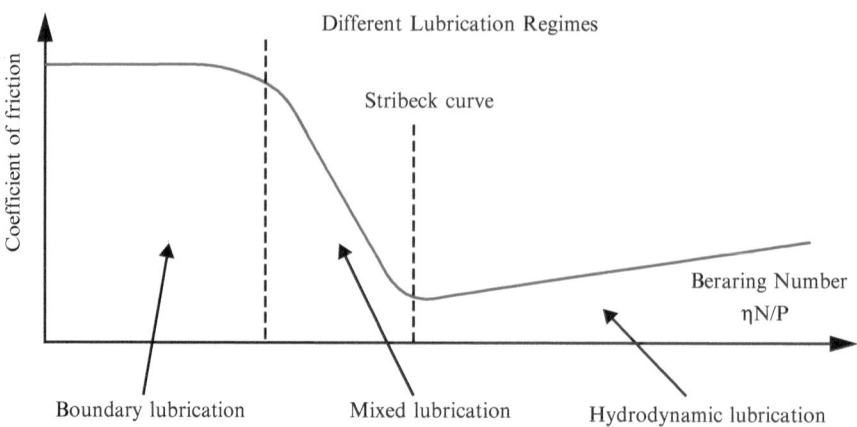

Fig. 2.8 Different regimes of lubrication

Four different forms of lubrication can be identified for self-pressure-generating lubricated contacts: (1) Hydrodynamic, (2) Elastohydrodynamic, (3) Partial or mixed, (4) Boundary. Different regimes of lubrication can be shown (Fig. 2.8) using Stribeck's curve (Stachowiak and Batchelor 1993) available in most literature dealing lubrication.

The Stribeck's curve presented is for a hypothetical fluid lubricated bearing system presents friction coefficient as a function of sliding speed, fluid viscosity and unit load. Three lubrication mechanisms; boundary, mixed, and hydrodynamic lubrications; have been marked on this plot. This plot defines the stability of lubrication. Say the operator is operating to the right of minimum friction and an increase in lubricant temperature happens which causes decrease in viscosity and hence a smaller value of bearing number. The coefficient of friction decreases, not as much as heat is generated in shearing the lubricant, and consequently lubricant temperature drops. Thus, the region to the right of the minimum defines stable lubrication because variations in this side are self-correcting. To the left of line, a decrease in viscosity would increase the friction. As temperature rise would increase, the viscosity would be reduced still more. The result would be compounded and the region represents unstable lubrication and load sharing.

2.4.1 *Hydrodynamic or Full Film Lubrication*

Hydrodynamic or full film lubrication is the condition when the load-carrying surfaces are separated by a relatively thick film of lubricant. This is a stable regime of lubrication and metal-to-metal contact does not occur during the steady state operation of the bearing. The lubricant pressure is self-generated by the moving surfaces drawing the lubricant into the wedge formed by the bounding surfaces at a high enough velocity to generate the pressure to completely separate the surfaces and support the applied load. The coefficient of friction is lower than with boundary-layer lubrication. In hydrodynamic lubrication, the following characteristics can be outlined:

1. With increase in load, the fluid film thickness at the point of minimum thickness decreases.
2. In this lubrication, the pressure within the fluid mass increases when the applied load decreases the film thickness.
3. Pressure within the fluid mass is highest at some point approaching minimum clearance and lowest at the point of maximum clearance.
4. Viscosity of the oil increases as the pressure increases. Also, film thickness increases with the use of more viscous oils.
5. Fluid friction increases as the viscosity of the lubricant becomes high.

2.4.2 Elastohydrodynamic Lubrication

Elastohydrodynamic lubrication is the condition that occurs when a lubricant is introduced between surfaces that are in rolling contact, such as ball and rolling element bearings. In this lubrication regime, the load is sufficiently high enough for the surfaces to elastically deform during the hydrodynamic action.

2.4.3 Partial or Mixed Lubrication

Partial or mixed lubrication regime deals with the condition when the speed is low, the load is high, or the temperature is sufficiently large to significantly reduce lubricant viscosity—when any of these conditions occur, the tallest asperities of the bounding surfaces will protrude through the film and occasionally come in contact.

2.4.4 Boundary Lubrication

Boundary Lubrication term was coined by English Biologist Sir Hardy in 1922. He quoted that "Very thin adsorbed layers, about 10 Å thick, were sufficient to cause two glass surfaces to slide over each other". The layer of lubricant separates sliding surfaces, i.e. no direct contact of the sliding parts. This situation is required for many applications, such as steel gears, piston-rings, and metal-working tools, to prevent severe wear or high coefficients of friction and seizure. The physical and chemical properties of thin surface films are of significant importance, while the properties of the bulk fluid lubricant are insignificant.

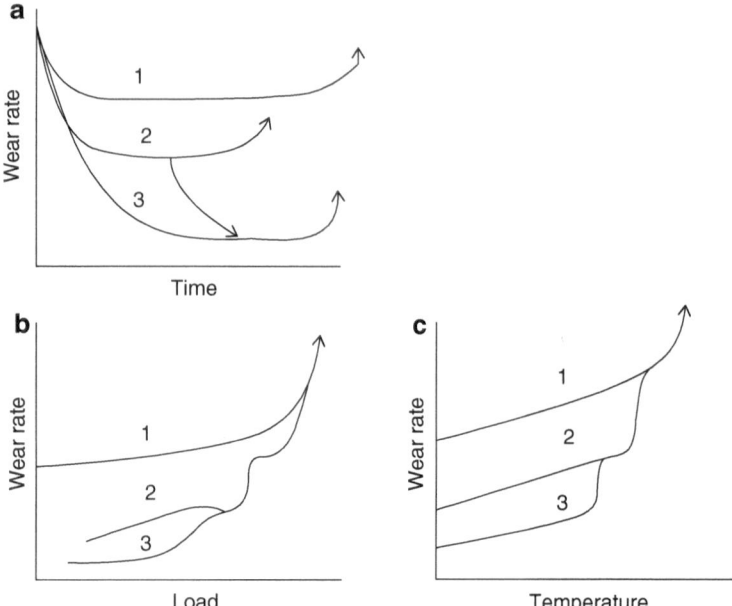

Fig. 2.9 Comparative study among dry (1), boundary (2), and hydrodynamic (3) lubrication mechanisms (**a**) Wear rate vs time, (**b**) Wear rate vs load, (**c**) Wear rate vs temperature (E-Learning courses from IITs/IISC NEPTEL)

2.4.5 Mechanisms of Boundary Lubrication

Physisorption: is the classical form of adsorption. Molecules of adsorbate may attach or detach from a surface without any irreversible changes to the surface or the adsorbate.

Chemisorption: It is an irreversible or partially irreversible form of adsorption which involves some degree of chemical bonding between adsorbate and substance.

  To understand lubrication in all respect, a comparative study among dry, boundary lubricated, and fluid film lubricated has been discussed here and is shown in Fig. 2.9 E-Learning courses from IITs/IISC NEPTEL. From the Fig. 2.9, one can observe that as the time increases, wear rate decreases and remain constant up to certain time, then increases for dry lubrication. For boundary lubrication, the wear rate decreases up to certain time, then decreases or increases depending on the improvement in surface smoothness. If surface smoothness occurs, boundary lubrication turns out to be fluid film lubrication which means wear rate decreases, otherwise wear rate increases. For fluid film lubrication, wear rate drastically decreases, then remain constant up to certain limit and then increases. Also, for dry lubrication, as the load increases wear rate increases. For boundary lubrication, wear rate increases and the rate of increase in wear rate is lesser than the dry lubrication. For fluid film lubrication also, wear rate increases, but rate of increase in wear rate is initially lower. Same process does occur with increase in the temperature. It should also be noticed that the wear rate increases for all three lubrication mechanisms with increase in temperature. One of the common elements in all machines is spur gear. Spur gear generally operates under boundary lubrication regime.

2.4.6 Hydrostatic Lubrication

In this mechanism of lubrication, the bearing surfaces are fully separated by a lubricating film of liquid or gas. The medium of lubrication is forced between the surfaces by an external source of pressure. A complete fluid film is present even at zero sliding speed as long as a continuous supply of pressurized lubricant is maintained. There is complete absence of sticking friction. Hydrostatic bearings can support very large masses and allow them to be moved from their stationary with the use of minimal force.

References

Chauhan A. Experimental and theoretical investigations of the thermal behaviour of some noncircular journal bearing profiles, Ph.D. thesis, Mechanical Engineering Department, NIT Hamirpur; 2011.

Chauhan A, Sehgal R. An experimentation investigation of the variation of oil temperatures in offset-halves journal bearing profile using different oils. Indian J Tribol. 2008;2:27–41.

E-Learning courses from IITs/IISC NEPTEL, Mechanical Engineering (Boundary Lubrication), Module 4. https://onlinecourses.nptel.ac.in/

Kumar R. Studies of the hydrodynamic bearings with surface profiling and entrained solid particulate, Industrial Tribology, Machine Dynamics and Maintenance Engineering Centre (ITMMEC), IIT Delhi; 2007.

Stachowiak GW, Batchelor AW. Engineering tribology. Amsterdam: Elsevier Science; 1993. p. 123–31.

Chapter 3
Performance Parameters

Nomenclature

b	Width of bearing, mm
C	Radial clearance, μm
C_h	Horizontal clearance for elliptical journal bearing, μm
C_m	Minimum clearance when journal centre is coincident with geometric centre of the bearing, μm
C_P	Specific heat of the lubricating oil, J/kg °C
ID	Inner diameter of the offset-halves journal bearing, mm
D_{Imin}	Minimum inner diameter of the elliptical journal bearing, mm
D_{Imax}	Maximum inner diameter of the elliptical journal bearing, mm
°C	Degree Celsius
e	Eccentricity, m
E_M	Elliptical ratio
h_c	Convection heat transfer coefficient of bush, W/m °C
h	Film thickness for offset-halves and elliptical journal bearing, mm
K_{oil}	Thermal conductivity of lubricating oil, W/m °C
K_s	Thermal conductivity of bearing, W/m °C
l	Length of the bearing, m
n	Number of iterations
N	Journal speed, rpm
ofr	Oil flow rate, lt/min
O_B	Bearing centre
O_J	Journal centre
O_L	Lower lobe centre
O_U	Upper lobe centre
OD	Outer diameter of the bearings, mm
P	Film pressure, Pa
$P(i,j)_{iso}$	Isothermal pressure, Pa

© The Author(s) - SpringerBriefs 2016
A. Chauhan, *Non-Circular Journal Bearings*, SpringerBriefs in Materials,
DOI 10.1007/978-3-319-27333-4_3

$P(i,j)_{th}$	Thermal pressure, Pa
PTPA	Parabolic temperature profile approximation
R	Radius of journal, mm
R_j	Radius of journal, mm
r	Bush radius, mm
Rbi	Inner bush radius, mm
R_L	Radius of lower lobe of the bearing, mm
R_U	Radius of upper lobe of the bearing, mm
RTD	Resistance temperature detector
s	Bearing surface
t	Thickness of bearing, m
T	Lubricating film temperature, °C
T_a	Ambient temperature, °C
T_b	Bush temperature, °C
T_L	Temperature of the lower bounding surface (journal), °C
T_m	Mean temperature across the film, °C
T_o	Oil inlet temperature, °C
T_s	Surface temperature, °C
T_U	Temperature of the upper bounding surface (bearing)
THD	Thermohydrodynamic
u, w	Velocity components in X- and Z-directions, m/s
u_L	Velocity of lower bounding surface, m/s
u_U	Velocity of upper bounding surface, m/s
U	Velocity of journal, m/s
x, y, z	Coordinates in circumferential, radial, and axial directions
φ	Attitude angle
ϕ_1, ϕ_2	Attitude angles from 0 to 180° (upper lobe) and 180–360° (lowerlobe), respectively
α	Barus viscosity-pressure index, Pa^{-1}
γ	Temperature-viscosity coefficient of lubricant, K^{-1}
δ	Offset factor (C_m/C)
ε	Eccentricity ratio
$\varepsilon_1, \varepsilon_2$	Eccentricity ratio from 0 to 180° (upper lobe) and 180–360° (lower lobe), respectively
θ	Angle measured from the horizontal split axis in the direction of rotation
μ	Absolute viscosity, Pa s
μ_{ref}	Absolute viscosity at oil inlet temperature, Pa s
ρ	Density of lubricating oil, kg/m^3
ω	Angular velocity of shaft, rad/s

3.1   Definitions

There may be number of parameters which can be explained in relation to non-circular journal bearings. Some of them related to thermal view has been listed and defined in the current section.

1. *Pressure*: Pressure developed inside the bearing represents the actual load a bearing can take. Because of variation in film thickness across the bearing profile, the pressure is developed in variable magnitude across the bearing. In non-circular journal bearing, an appreciable positive pressure is observed in lower lobe and upper lobe, which results in better stability of such kind of journal-bearing profiles.

2. *Temperature*: During operation of any bearing, the temperature of lubricant as well as bush rises to certain level. The temperature rise is observed in both lobes of the non-circular journal bearing like pressure; however, a comparatively less rise is observed when compared with circular journal bearings of the same dimensions and operating at the same operating conditions.

3. *Load capacity*: It represents how much a journal bearing can support at given operating conditions before failure. As defined earlier, the pressure rise in the bearing decides how much a bearing can withstand; the load capacity can be found out from the pressure profile obtained using Simpson rule as explained in Sect. 3.2.5.

4. *Sommerfeld number*: In the design of fluid film bearings, Sommerfeld number is a dimensionless quantity used extensively in hydrodynamic lubrication analysis. The Sommerfeld number is very important in lubrication analysis because it contains all the variables normally specified by the designer. The Sommerfeld number is named after Arnold Sommerfeld (1868–1951).

5. *Power loss*: The power loss in a bearing happens as a result of bearing friction. Power losses have been evaluated by determination of shear forces, and then employing the Simpson's 1/3rd rule (also explained in Sect. 3.2.5 in this chapter).

6. *Dynamic coefficient*: Though the book aims at thermal study of non-circular journal bearings, an introduction to dynamic coefficient has also been given for the reader knowledge. Dynamic coefficients play a very important role for the stability of hydrodynamic journal bearings and thus are useful for the design of such bearings. Therefore, it is important to determine the stiffness, damping, and added mass coefficients of the hydrodynamic bearing.

7. *Whirl stability*: The various factors which may lead to whirling in Turbomachinery are: internal friction damping, hydrodynamic fluid film bearings and seals, aerodynamic cross-coupling forces, dry friction wheel, asymmetric shaft properties, entrained fluid in rotor, gyroscopic-induced whirling, electromagnetic forces, and many more Edgar (1972). The whirling in fluid film bearing is known as oil whip or half frequency whirl and large unbalance may suppress the whirl motion in such type of bearings. The non-circular journal bearings operate with two pressure zones: one on lower portion (called as lower lobe) and second is on upper portion (called as upper lobe). Therefore, it helps in stability of rotor and it has also been reported by San Andres (2006) that most of the non-circular journal bearing have the excellent suppression of whirl at high speeds; however, may be subject to whirl problem at high speeds. He also suggested that the non-circular bearings can be used at low or moderate speeds.

8. *Cavitation*: For a good precision and high value of load on the bearings, the effect of cavitation may be of great interest. In a journal bearing, a very thin oil film is formed in the passage between the journal and housing. Due to the eccentricity of the gap between journal and housing, the thickness of oil film is not uniform. Where the oil film is thin, a low pressure region is formed, and if the pressure drops below the vaporization pressure, cavitation occurs. Cavitation may occur when the dissolved gas comes out of solution as the pressure falls in the diverging section of the bearing. Further, the point to be kept in mind is that lubricant itself may evaporate, if the pressure drop is low enough (below the vapor pressure). The cavitation can lead to surface pitting by bubble collapse and it can significantly affect the bearing forces, and even can also alter the orientation of shaft. During cavitation, oil starts to cavitate in the low pressure region just downstream of the location of minimum gap and is independent of the shaft position. The Elrod's algorithm method is simple to use and automatically implements cavitation boundary conditions at film rupture and reformation. Another significant aspect of the film profile is the presence of entrained air or vapor bubbles as the bubbly oil is believed to have an increased load-carrying capacity by enhancing the oil viscosity. It has been observed that when the bearing is subjected to dynamic squeeze action, the presence of bubbles cans significantly affects the load-carrying capacity of the bearing (Kasolang and Dwyer-Joyce 2008).

3.2   Mathematical Analysis

The governing equations as well as the numerical procedures adopted for the mathematical modeling of non-circular journal bearing have been presented in this chapter. In the simplification of energy equation, people may adopt a number of techniques available in literature; however, the parabolic temperature profile approximation (PTPA) across the film thickness has been presented here.

3.2.1   Film Thickness Equations

Though there are different types of non-circular journal bearings, solution procedure for some of them is discussed here and the same steps can be repeated after achieving the film thickness for configuration other than shown here. The film thickness equation for a circular journal bearing is given as:

$$h = C(1 + \varepsilon \cos\theta) \tag{3.1}$$

In (3.1), h represents film thickness for circular journal bearing, C represents radial clearance, ε represents eccentricity ratio, and θ represents angle measured from the horizontal split axis in the direction of rotation.

The film thickness equations for offset-halves journal bearings are given as (Sehgal et al. 2000):

$$h = c_m \left[\left(\frac{1+\delta}{2\delta} \right) + \left(\frac{1-\delta}{2\delta} \right) \cos\theta - \varepsilon \sin(\phi - \theta) \right] \quad (0 < \theta < 180) \quad (3.2)$$

$$h = c_m \left[\left(\frac{1+\delta}{2\delta} \right) - \left(\frac{1-\delta}{2\delta} \right) \cos\theta - \varepsilon \sin(\phi - \theta) \right] \quad (180 < \theta < 360) \quad (3.3)$$

In the subsequent sections, various bearing performance parameters have been obtained while using (3.2) and (3.3) for upper lobe and lower lobe of the bearing, respectively. In these equations, C_m denotes minimum clearance when journal centre is coincident with geometric centre of the bearing, δ denotes offset factor (C_m/C), and ϕ denotes attitude angle.

The film thickness equations for elliptical journal bearing have been adopted from Hussain et al. (1996) and are given as:

$$h = c_m \left[1 + E_M + \varepsilon_1 \cos(\theta + \phi - \phi_1) \right], \quad \text{for } 0 < \theta < 180 \quad (3.4)$$

$$h = c_m \left[1 + E_M + \varepsilon_2 \cos(\theta + \phi - \phi_2) \right], \quad \text{for } 180 < \theta < 360 \quad (3.5)$$

Different parameters in (3.4) and (3.5) are given as:

$$\varepsilon_1 = \left(E_M^2 + \varepsilon^2 - 2E_M \varepsilon \cos\phi \right)^{\frac{1}{2}}; \quad \varepsilon_2 = \left(E_M^2 + \varepsilon^2 + 2E_M \varepsilon \cos\phi \right)^{\frac{1}{2}}$$

$$\phi_1 = \pi - \tan^{-1}\left(\frac{\varepsilon \sin\phi}{E_M - \varepsilon \cos\phi} \right); \quad \phi_2 = \tan^{-1}\left(\frac{\varepsilon \sin\phi}{E_M + \varepsilon \cos\phi} \right); \quad E_M = \left(\frac{C_h - C_m}{C_m} \right)$$

In (3.4) and (3.5), h represents film thickness for elliptical journal bearing, E_M represents elliptical ratio, $\varepsilon_1, \varepsilon_2$ represents eccentricity ratio from 0 to 180° (upper lobe) and 180–360° (lower lobe), respectively, ϕ_1, ϕ_2 represents attitude angles from 0 to 180° (upper lobe) and 180–360° (lower lobe), respectively, and C_h represents horizontal clearance for elliptical journal bearing.

3.2.2   Reynolds Equation

The phenomenon of hydrodynamic lubrication can be expressed mathematically in the form of an equation which was originally derived by Reynolds and is commonly known throughout the literature as the 'Reynolds equation'. It can be derived from the full solution of Navier–Stokes momentum and continuity equation since it is a simplification of this or by simply considering the equilibrium of an element of liquid subjected to viscous shear and applying the continuity of flow principle.

All the simplifying assumptions necessary for the derivation of the Reynolds equation are listed below (Stachowiak and Batchelor 1993):

1. Body forces are neglected, i.e. there are no extra outside fields of forces acting on the fluids.
2. Pressure is constant through the film.
3. No slip at the boundaries as the velocity of the oil layer adjacent to the boundary is the same as that of the boundary.
4. Flow is laminar and viscous.
5. Lubricant behaves as a Newtonian fluid.
6. Inertia and body forces are negligible compared with the pressure and viscous terms.
7. Fluid density is constant. Usually valid for fluids, when there is not much thermal expansion.
8. There is no vertical flow across the film.

For steady state and incompressible flow, the Reynolds equation is (Hussain et al. 1996):

$$\frac{\partial}{\partial x}\left(\frac{h^3}{\mu}\frac{\partial p}{\partial x}\right) + \frac{\partial}{\partial z}\left(\frac{h^3}{\mu}\frac{\partial p}{\partial z}\right) = 6U\frac{\partial h}{\partial x} \qquad (3.6)$$

Here, P represents film pressure, μ represents absolute viscosity of the lubricant, and U represents velocity of journal. Equation (3.6) can be set into finite differences by using central difference scheme. During the solution, the effect of viscosity variation due to temperature rise during operation must be considered. The variation of viscosity with temperature and pressure can be simulated using the following viscosity relation:

$$\mu = \mu_{\text{ref}} e^{\alpha P - \gamma(T - T_0)} \qquad (3.7)$$

In (3.7), μ_{ref} represents absolute viscosity of the lubricant at oil inlet temperature, γ represents temperature–viscosity coefficient of lubricant, α represents Barus viscosity–pressure index, T represents lubricating film temperature, and T_0 represents oil inlet temperature.

3.2.3   Energy Equation

The temperature distribution within the lubricant film is determined by the solution of the energy equation and satisfying appropriate boundary conditions mentioned under the section of boundary conditions. The energy equation for steady state and incompressible flow is given as (Sharma and Pandey 2007):

$$\rho C_p\left(u\frac{\partial T}{\partial x} + w\frac{\partial T}{\partial z}\right) = \frac{\partial}{\partial y}\left(K\frac{\partial T}{\partial y}\right) + \mu\left[\left(\frac{\partial u}{\partial y}\right)^2 + \left(\frac{\partial w}{\partial y}\right)^2\right] \qquad (3.8)$$

Here, C_P represents specific heat of the lubricating oil, K represents thermal conductivity of the lubricating oil, and u, w represents velocity components in X- and Z-directions. In (3.8), the terms due to conduction and convection have been omitted as the magnitude of these terms is usually very small in comparison to the convection along the film and conduction across the film. In (3.8), the left-hand side represents energy transfer due to convection, the first term on right hand side represents energy transfer due to conduction, and the second term on right hand side represents energy transfer due to dissipation. In this equation, x-axis represents axis along the circumference of the bearing, whereas y-axis and z-axis have been considered along the oil-film thickness and across the width of bearing, respectively. Equation (3.8) can be solved directly by coupling Reynolds and energy equation or by PTPA techniques, even currently some researchers have reported CFD-based analysis for such kind of journal bearings.

3.2.3.1 Formulation of energy equation with PTPA

The variation of temperature across the film thickness in energy equation is approximated by parabolic temperature profile. The expressions for velocities appearing in energy equation are obtained by double integration of momentum equations, $\dfrac{\partial P}{\partial x} = \dfrac{\partial}{\partial y}\left(\mu\dfrac{\partial u}{\partial y}\right)$ and $\dfrac{\partial P}{\partial z} = \dfrac{\partial}{\partial y}\left(\mu\dfrac{\partial w}{\partial y}\right)$, across the film thickness using appropriate boundary conditions (at $y = 0$; $u = u_{\mathrm{L}}$, $w = 0$ and, at $y = h$; $u = u_{\mathrm{U}}$, $w = 0$) as:

$$u = \left(\frac{y^2 - yh}{2\mu}\right)\frac{\partial P}{\partial x} + u_{\mathrm{L}}\left(\frac{h-y}{h}\right) + u_{\mathrm{U}}\left(\frac{y}{h}\right) \tag{3.9}$$

$$w = \left(\frac{y^2 - yh}{2\mu}\right)\frac{\partial P}{\partial z} \tag{3.10}$$

where, u_{L} and u_{U} represent the velocity of lower and upper bounding surfaces, respectively. Equations (3.6), (3.9), and (3.10) may also be used only as approximation for non-viscous flow if viscosity in (3.6) is calculated using a well chosen, height-averaged, temperature. The temperature profile across the film thickness is represented by a second-order polynomial as:

$$T = a_1 + a_2 y + a_3 y^2 \tag{3.11}$$

In order to evaluate the constants appearing in (3.11), the following boundary conditions are used:

$$\text{At } y = 0, \quad T = T_{\mathrm{L}} \tag{3.11a}$$

$$\text{At } y = h, \quad T = T_{\mathrm{U}} \tag{3.11b}$$

$$\text{and } T_{\mathrm{m}} = \frac{1}{h}\int_0^h T\,dy \qquad\qquad (3.11c)$$

Thus, temperature profile expression (written in (3.11)) takes the following form:

$$T = T_{\mathrm{L}} - \left(4T_{\mathrm{L}} + 2T_{\mathrm{U}} - 6T_{\mathrm{m}}\right)\left(\frac{y}{h}\right) + \left(3T_{\mathrm{L}} + 3T_{\mathrm{U}} - 6T_{\mathrm{m}}\right)\left(\frac{y}{h}\right)^2 \qquad (3.12)$$

where, T_{L}, T_{U}, and T_{m} represent temperatures of the lower bounding surface (journal), upper bounding surface (bearing), and mean temperature across the film, respectively (Fig. 3.1).

$$-k_s\left(\frac{\partial T_s}{\partial x_s}\right)_{x_s=l} = h_c\left(T_s\left(l, y_s\right) - T_a\right)$$

$$k_{\mathrm{oil}}\left(\frac{\partial T}{\partial y}\right)_{\substack{\text{upper}\\\text{bounding}\\\text{surface}}} = k_s\left(\frac{\partial T_s}{\partial y_s}\right)_{y_s=0}$$

$$-k_s\left(\frac{\partial T_s}{\partial y_s}\right)_{y_s=t} = h_c\left(T_s\left(x_s, t\right) - T_a\right)$$

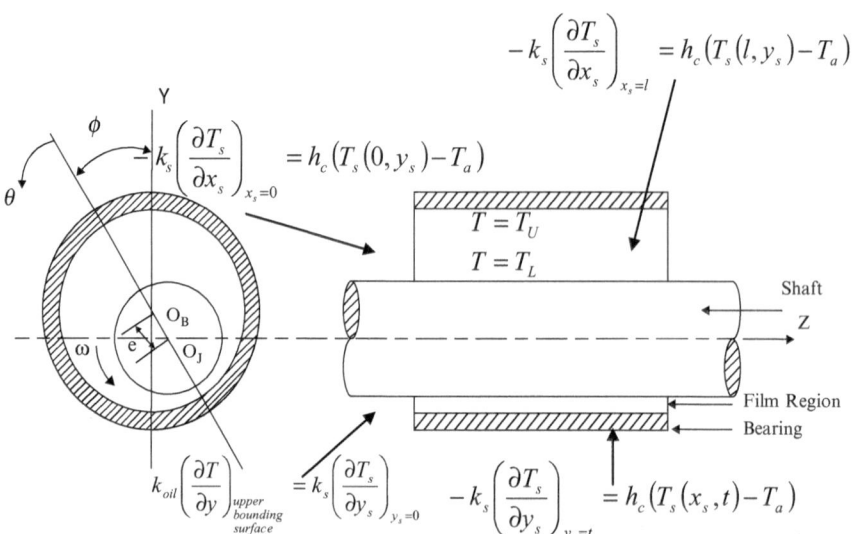

Fig. 3.1 Bearing geometry with coordinate system

$$-k_s \left(\frac{\partial T_s}{\partial x_s} \right)_{x_s=0} = h_c \left(T_s \left(0, y_s \right) - T_a \right)$$

$$T = T_U$$

$$T = T_L$$

Substitution of 'u', 'w', and 'T' expressions ((3.9), (3.10), and (3.12)) into energy equation (3.8) and subsequently integrating the energy equation across the film thickness from the limit '0' to 'h' yield the following form of energy equation.

$$6T_L + 6T_U - 12T_m - \frac{\rho C_p h^4}{120 K \mu} \frac{\partial P}{\partial x} \left(\frac{\partial T_L}{\partial x} + \frac{\partial T_U}{\partial x} - 12 \frac{\partial T_m}{\partial x} \right) - \frac{\rho C_p h^4}{120 K \mu} \frac{\partial P}{\partial z}$$
$$\times \left(\frac{\partial T_L}{\partial z} + \frac{\partial T_U}{\partial z} - 12 \frac{\partial T_m}{\partial z} \right) - \frac{\rho C_p h^2 \left(u_L + u_U \right)}{2K} \frac{\partial T_m}{\partial x} - \frac{\rho C_p h^2 \left(u_U - u_L \right)}{12K} \left(\frac{\partial T_U}{\partial x} - \frac{\partial T_L}{\partial x} \right) \quad (3.13)$$
$$+ \frac{h^4}{12 K \mu} \left[\left(\frac{\partial P}{\partial x} \right)^2 + \left(\frac{\partial P}{\partial z} \right)^2 \right] + \frac{\mu \left(u_U - u_L \right)^2}{K} = 0$$

3.2.4   Heat Conduction Equation

The temperature in bush is determined by using the Laplace equation within the bearing material as given below (Hori 2006):

$$\frac{\partial^2 T_b}{\partial x^2} + \frac{\partial^2 T_b}{\partial y^2} + \frac{\partial^2 T_b}{\partial z^2} = 0 \quad \left(\text{In Cartesian coordinate} \right) \qquad (3.14)$$

In this equation, T_b stands for bush temperature. Equation (3.14) can be then set into finite differences by using central difference technique.

The boundary conditions used in the solution of governing equations are:

$$P = 0 \text{ at } x = 0 \quad \text{and} \quad x = l$$

$$u = u_L \text{ at } y = 0 \quad \text{and} \quad 0 \le x \le l$$

$$u = 0 \text{ at } y = h \quad \text{and} \quad 0 \le x \le l$$

$$T = T_0 \text{ at } x = 0 \quad \text{and} \quad 0 \le y \le h$$

$$T = T_L \text{ at } y = 0 \quad \text{and} \quad 0 \le x \le l$$

$$T = T_U \text{ at } y = h \quad \text{and} \quad 0 \le x \le l$$

$$T(0,y) = T_0; \quad T(x,0) = T_0; \quad k_{oil}\left(\frac{\partial T}{\partial y}\right)_{\substack{upper \\ bounding \\ surface}} = k_s\left(\frac{\partial T_s}{\partial y_s}\right)_{y_s = 0}$$

$$-k_s\left(\frac{\partial T_s}{\partial y_s}\right)_{y_s = t} = h_c\left(T_s(x_s,t) - T_a\right); \quad -k_s\left(\frac{\partial T_s}{\partial x_s}\right)_{x_s = 0} = h_c\left(T_s(0,y_s) - T_a\right)$$

$$-k_s\left(\frac{\partial T_s}{\partial x_s}\right)_{x_s = l} = h_c\left(T_s(l,y_s) - T_a\right); \quad k_s\left(\frac{\partial T_s}{\partial z_s}\right)_{z_s = 0} = h_c\left(T_s(x_s,,y_s,,0) - T_a\right)$$

$$-k_s\left(\frac{\partial T_s}{\partial z_s}\right)_{z_s = b} = h_c\left(T_s(x_s,,y_s,,b) - T_a\right)$$

where, K_s denotes thermal conductivity of bearing, h_c denotes convection heat transfer coefficient of bush, l denotes length of the bearing, s denotes bearing surface, t denotes thickness of bearing, b denotes width of bearing, and T_a ambient temperature.

The above governing equations can be solved by satisfying the following convergence criterions:

For pressure:

$$\frac{\left|\left(\sum P_{i,j}\right)_{n-1} - \left(\sum P_{i,j}\right)_n\right|}{\left|\left(\sum P_{i,j}\right)_n\right|} \leq 0.0001 \tag{3.15}$$

For temperature:

$$\frac{\left|\left(\sum T_{i,j}\right)_{n-1} - \left(\sum T_{i,j}\right)_n\right|}{\left|\left(\sum T_{i,j}\right)_n\right|} \leq 0.0001 \tag{3.16}$$

where, n represents number of iterations.

3.2.5   Cavitation

The effect of cavitation on the performance of any bearing can be studied using any of the methods discussed in methods of analysis. However, while studying the effect of cavitation, our focus must be on negative pressure which normally has been taken zero in most of the reported analysis in the literature.

3.2.6 Computational Procedure

The theoretical solutions of Reynolds, energy, and heat conduction equations can be adopted for all kinds of journal bearing. The temperature of upper and lower bounding surfaces may be assumed constant throughout and can be taken equal to the oil inlet temperature for first iteration. For subsequent iterations, the temperatures at oil bush interface are computed using heat conduction equation and appropriate boundary conditions. The numerical procedure adopted for obtaining the thermohydrodynamic solution is discussed below.

A suitable initial value of attitude angle can be assumed and then solve the film thickness equations. Use over-relaxation method with relaxation factor of 1.7 for error convergence. The initial viscosity values are assumed to be equal to the inlet oil viscosity. The numerical solution (by Finite Difference Method) begins with the known pressure distributions obtained by solution of Reynolds equation in above section. Viscosity in the fluid film domain corresponding to computed temperatures and pressures is calculated using (3.7). With new value of viscosity, i.e., after incorporating thermal effects, (3.6) has been solved for thermal pressure. With new value of pressure and viscosity, energy equation (3.13) has been further solved. Mean temperatures obtained by solving (3.13) are substituted in (3.12) to find the temperature profile in the oil film. Now, this temperature is used to solve (3.14) to obtain the temperature variation in the bush. The computation is continued till converged solutions for thermal pressure loop and temperature loop have been arrived. Under-relaxation with relaxation factor of 0.7 has been used for error convergence. The load-carrying capacity is obtained by applying the Simpson's 1/3rd rule to the pressure distribution. In computation, wherever reverse flow arises in domain, upwind differencing has been resorted to. Power losses have been evaluated by determination of shear forces, and then employing the Simpson's 1/3rd rule.

3.3 Experimental Procedure

Experimental work provides a greater insight into the performance of bearings under actual operating conditions. The test rigs to conduct thermal investigations on non-circular journal bearings have been elaborated here.

3.3.1 Description of Test Rig

A schematic diagram of the journal-bearing test rig used for experimental studies is shown in Fig. 3.2, whereas Fig. 3.3 shows the pictorial view of one of the non-circular journal bearings named: offset-halves journal bearing. The temperature is measured at 22 different points on circumference. The test rig has provision to

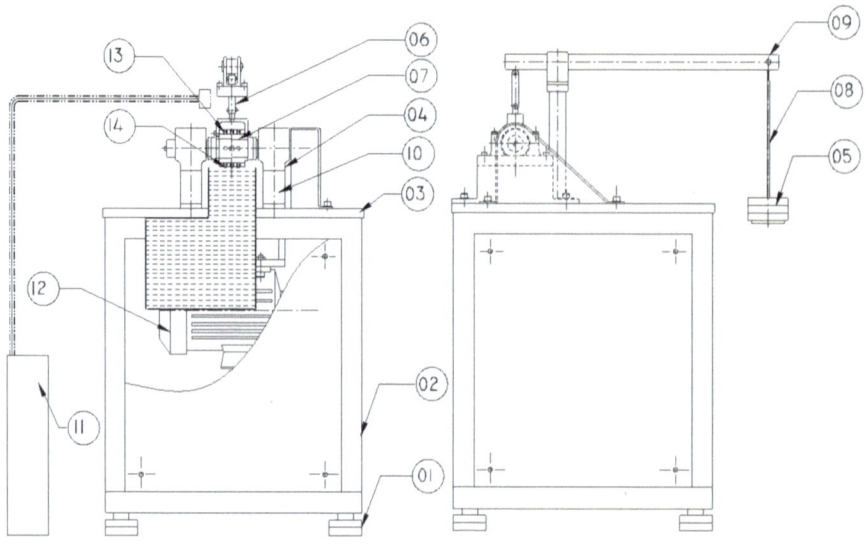

Parts	Description		
01	Antivibration pad	08	Loading pan
02	Structure	09	Lever
03	Base plate	10	Time belt
04	Plummer block	11	Lube system
05	Dead weight	12	Motor
06	Lifting support lever	13	Journal bearing
07	Sleeve	14	Journal bearing

Fig. 3.2 Journal-bearing test rig (Chauhan 2011)

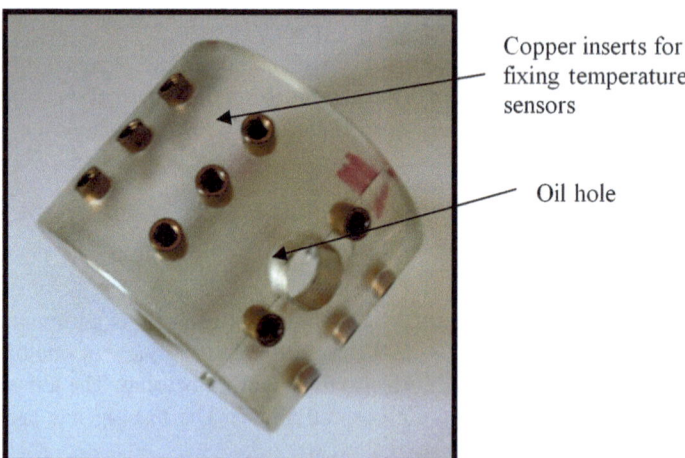

Copper inserts for fixing temperature sensors

Oil hole

Fig. 3.3 Offset-halves test bearing

measure temperature on middle and ends (on three circumferential planes) at every 45° angular position on the circumference of bearing (bush). The journal is made of C45 steel material and is mounted horizontally on two pedestal bearings. The journal is rotated by a motor through 1:2 ratio pulleys, to attain speed up to 5700 rpm. The motor speed is varied by a variable frequency drive; the drive frequency is changed by a potentiometer knob provided on controller front panel. A chrome-plated journal sleeve is tightened at middle portion of journal with lock nuts and bearing slides over it.

The bearing (bush) is made of Methyl Methacrylite with inner diameter=65 mm, outer diameter=85 mm and length=65 mm. On the circumference of bearing, copper inserts are fixed at 22 locations for measuring circumferential temperature. Figure 3.4a, b shows the locations and mounting of RTD probes for temperature

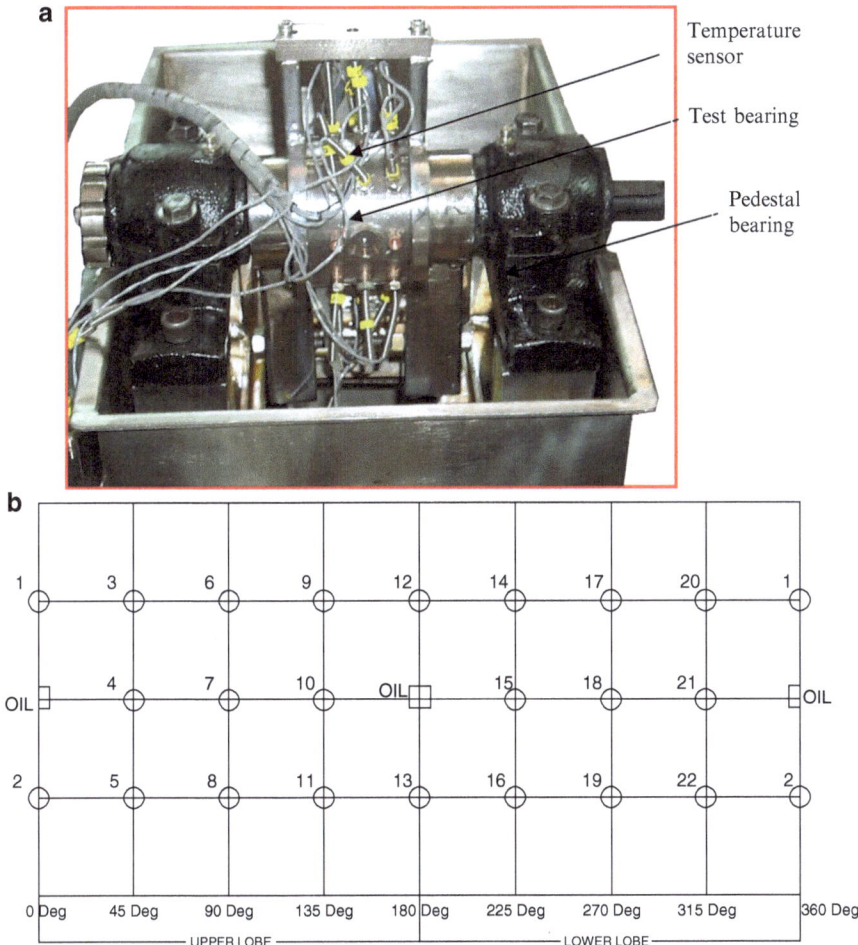

Fig. 3.4 (**a**) Test bearing with temperature sensors. (**b**) Diagram showing locations of mounting of RTD sensors (1–22) for temperature measurement

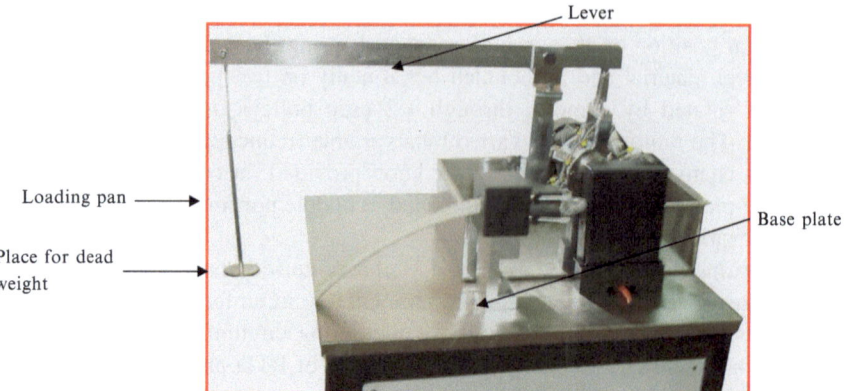

Fig. 3.5 Loading arrangement on the bearing in test rig

measurement. The inner side is open to oil circulating inside the journal and bearing interface, and thus sensors have direct access to the circulating oil. Apart from this, one additional sensor is fixed to measure the inlet temperature of oil. A sleeve fits over journal and radial load is applied on bearing by 1:5 loading lever. Smaller end of the lever is hooked to the bearing by link and on bigger end a loading pan is suspended. The ends of sensor are terminated on the metal box, from there cables carry signal to the controller. Under the load, when oil temperature increases, it is acquired and displayed on personnel computer (PC).

Radial load is applied on bearing by placing dead weights on the loading pan (Fig. 3.5). Radial load and journal speed can be varied to suit the test conditions. The test rig is designed to apply a maximum radial load up to 700 N.

An rpm sensor disc (Fig. 3.6) with square shape is fixed to journal end and rotates along with it. Proximity sensor is mounted perpendicular to it on bracket fixed to the base plate; signal is generated when sensor disc approached the active surface within the specified switching distance. This sensor functions in contact-less fashion and does not require any sensing mechanisms. An inductive proximity sensor is used as it has an excellent means of detecting the presence of a wide range of metallic targets. This detection is accomplished without contacting the target and is mechanically wear-free. It is comprised of a high frequency oscillator circuit followed by level detector, a post-amplification signal circuit, and drives a buffered solid-state output. When sensor disc is brought within the effective range of emitted field of the oscillator, a damping action results, which reduces the amplitude of oscillator. This amplitude shift is converted to digital signal by the level detector, which drives a buffer stage. When the object is removed, the oscillator and digital output is turned to its former state. The controller is housed inside the metal box.

A power ON switch is fixed on front panel for switching ON power supply, a potentiometer knob for regulating speed, and push button TEST ON and OFF switch

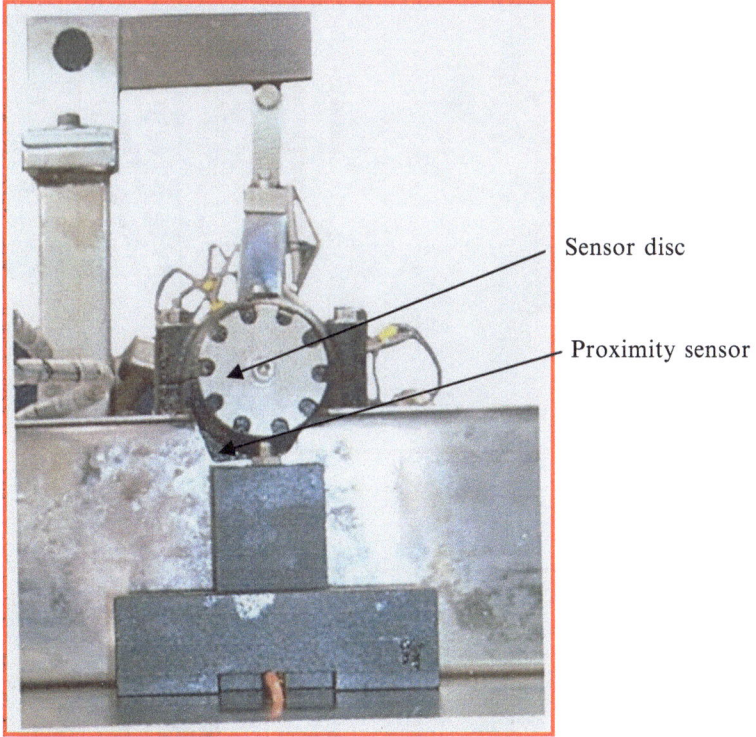

Fig. 3.6 An rpm sensor disc

is provided to begin and end the test. On back panel, a 2 A fuse holder, a 19 pin MS connector for control cables, 25 pin D connector for signal output 1 and 25 pin D connector for signal output 2 and 19 pin D connector for signal input are provided. Data acquisition system includes signal output cable 1 and 2, and WINDUCOM 2005 software. The measured data of oil temperature and speed obtained during the experimentation in analog form is converted into digital form and preprocessed on the instrumentation card. Then this data is serially transmitted to PC through signal output cable 1 and 2. The data is received, displayed, and stored on the PC using Lab view-based software WINDUCOM 2005.

Lubrication unit is made of a metallic tank with a motor and pump, by pass valve, control valve, pressure gauges, flow meter, inlet, and delivery pipe (Fig. 3.7). The discharge of the oil is controlled by means of gate valves provided. Two oil inlets on journal bearing are provided with a pipe, its flow is regulated by two gate valves, and pressure at that point is indicated by two pressure gauges fixed in line. An oil sump is provided beneath the bearing for collecting the used oil, which flows into metallic tank for its re-circulation. Figure 3.8 shows the main structure of the journal-bearing test rig.

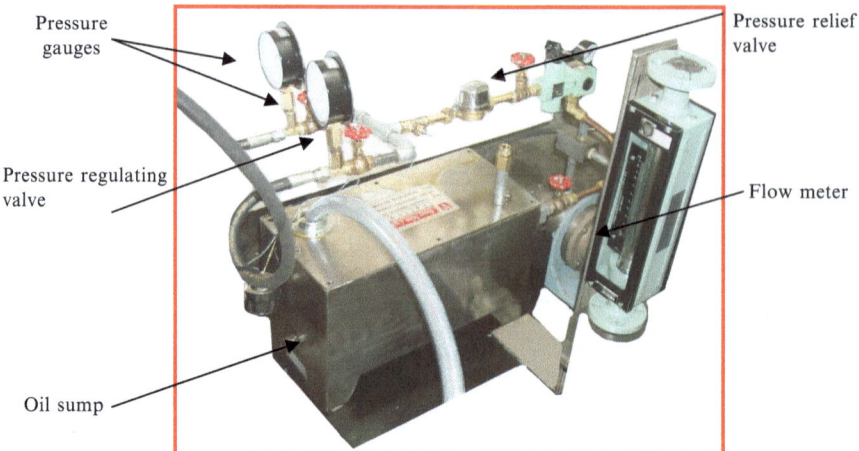

Fig. 3.7 Lubrication unit

Fig. 3.8 Main structure of
the test rig (Chauhan 2011)

Fig. 3.9 Main structure of
journal-bearing test rig

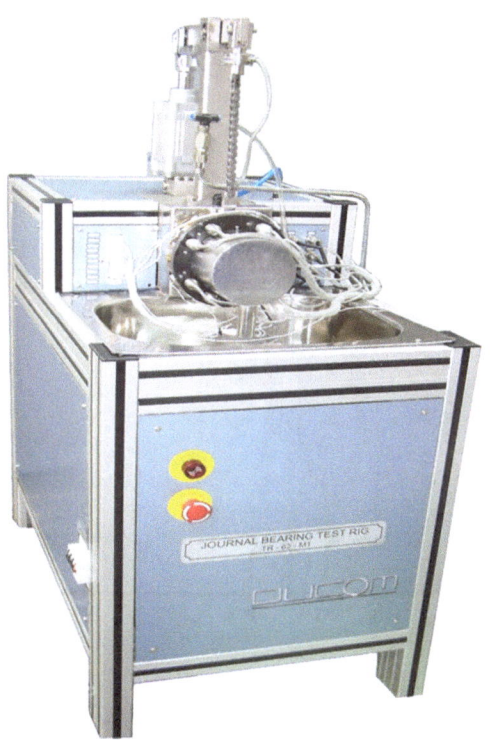

The test rig developed has been further modified to take higher loads and new test rig has been developed under AICTE Research Project (letter No. 20/AICTE/ RIFD/RPS(POLICY-II)77/2012-13) and is also compact in design. The new load capacity is 2000 N and speed at which the test rig can run is 6000 rpm. The load arrangement has been changed to pneumatic loading. Further, the test rig is now capable of measuring pressure developed in addition to the measurement of oil temperature rise during the operation. Figure 3.9 shows the new test developed and Fig. 3.10 represents the pressure and temperature sensors.

The lubricant supply is made using the lubricant unit as shown in Fig. 3.11. The test rig is equipped with cooling of lubricant during operation time and helps in keeping the lubricant temperature minimal, making it suitable for re-circulation.

Example: The performance parameters like pressure, oil-film temperature, Sommerfeld number, power loss and load capacity for an orthogonal journal bearing computed through computer programming, and oil-film pressure variation obtained experimentally have been given below (Figs. 3.12, 3.13, 3.14, 3.15, 3.16, and 3.17).

Fig. 3.10 Locations temperature and pressure sensors and loading lever

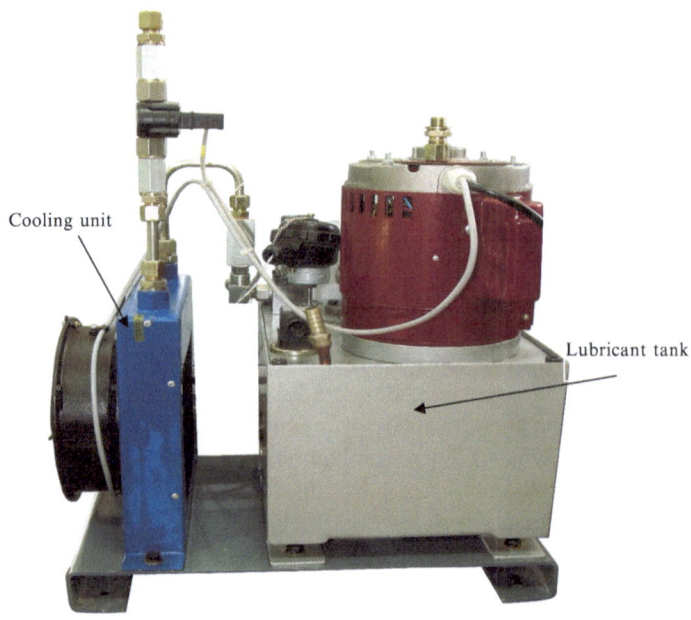

Fig. 3.11 Lubrication tank with cooling unit

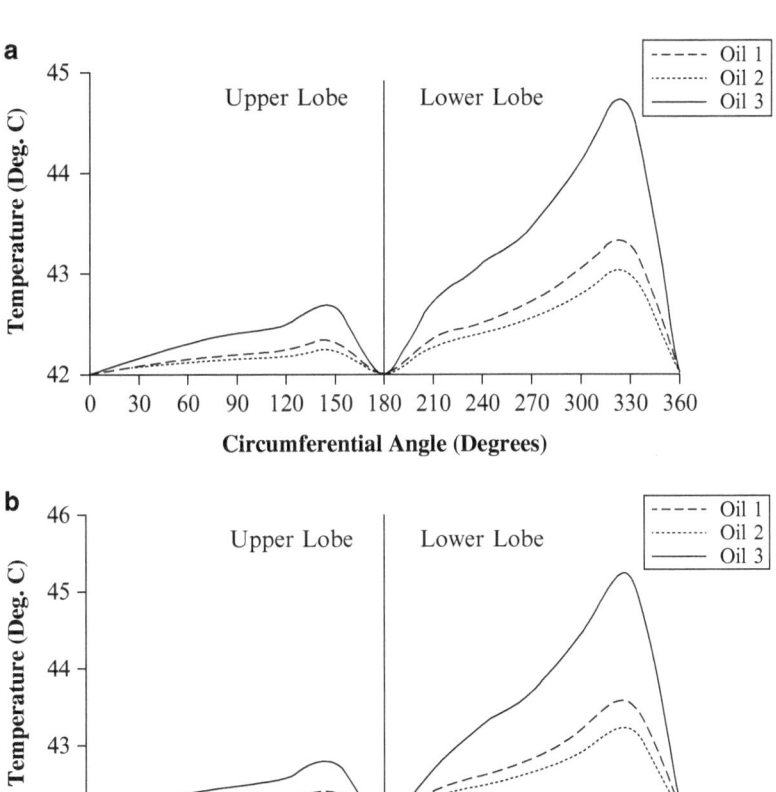

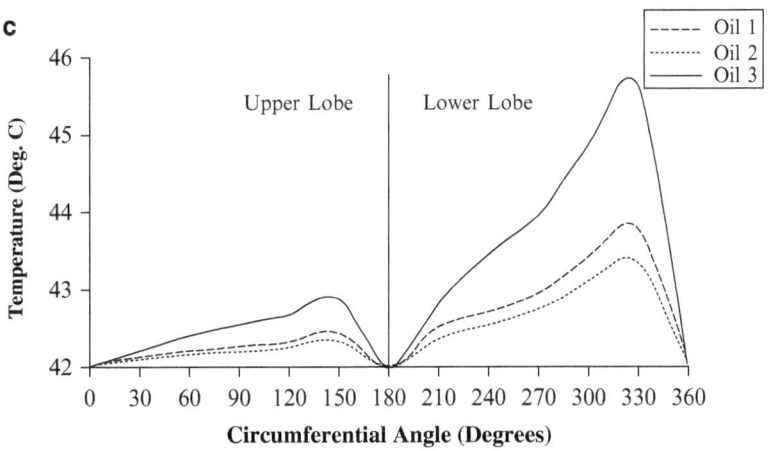

Fig. 3.12 (**a**) Variation of oil film temperatures in the central plane of the offset-halves bearing for different grade oils at 3000 rpm, oil inlet temperature = 42 °C and eccentricity ratio = 0.6. (**b**) Variation of oil film temperatures in the central plane of the offset-halves bearing for different grade oils at 3500 rpm, oil inlet temperature = 42 °C and eccentricity ratio = 0.6. (**c**) Variation of oil film temperatures in the central of the offset-halves bearing for different grade oils at 4000 rpm, oil inlet temperature = 42 °C and eccentricity ratio = 0.6

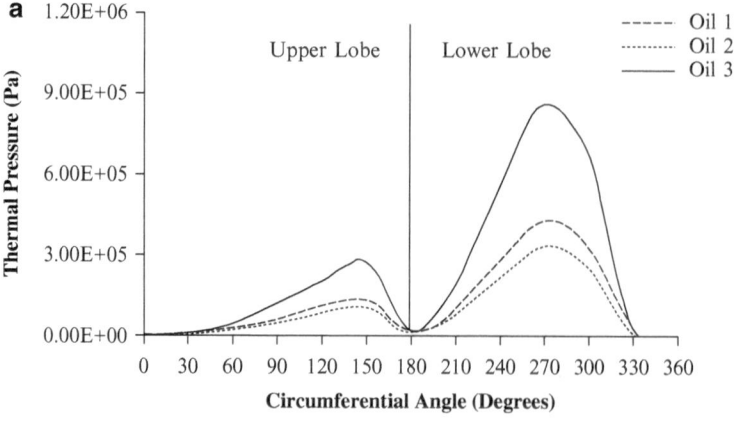

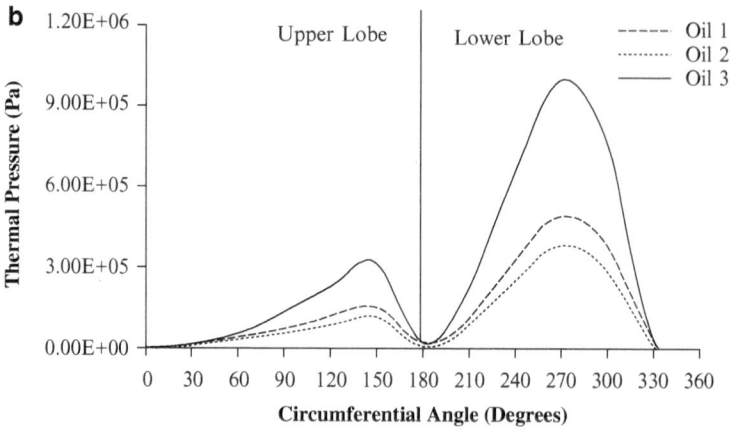

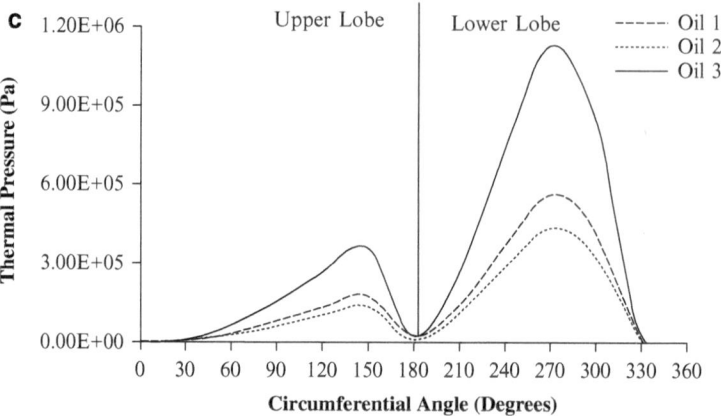

Fig. 3.13 (**a**) Variation of thermal pressure in the central plane of the offset-halves bearing for different grade oils at 3000 rpm, oil inlet temperature=42 °C and eccentricity ratio=0.6. (**b**) Variation of thermal pressure in the central plane of the offset-halves bearing for different grade oils at 3500 rpm, oil inlet temperature=42 °C and eccentricity ratio=0.6. (**c**) Variation of thermal pressure in the central plane of the offset-halves bearing for different grade oils at 4000 rpm, oil inlet temperature=42 °C and eccentricity ratio=0.6

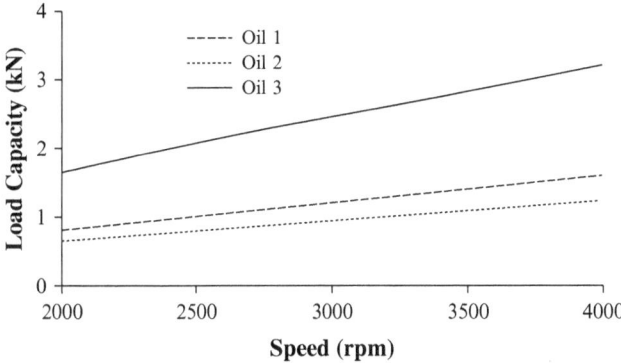

Fig. 3.14 Variation of the load capacity of offset-halves bearing with speed for all oils under study at oil inlet temperature=42 °C and eccentricity ratio=0.6

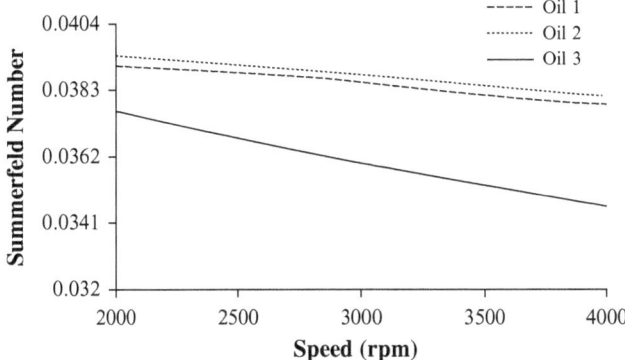

Fig. 3.15 Variation of the Sommerfeld number of offset-halves bearing with speed for all oils under study at oil inlet temperature=33 °C and eccentricity ratio=0.6

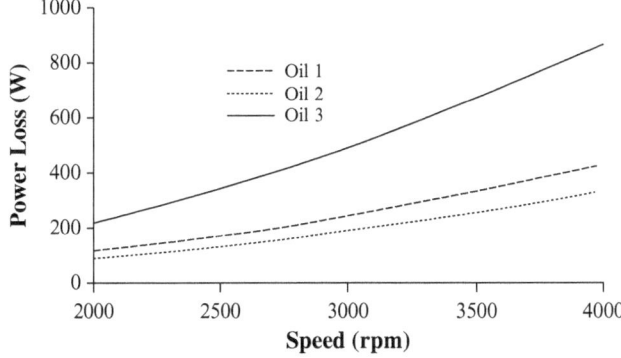

Fig. 3.16 Variation of the power loss of offset-halves bearing with speed for all oils under study at oil inlet temperature=42 °C and eccentricity ratio=0.6

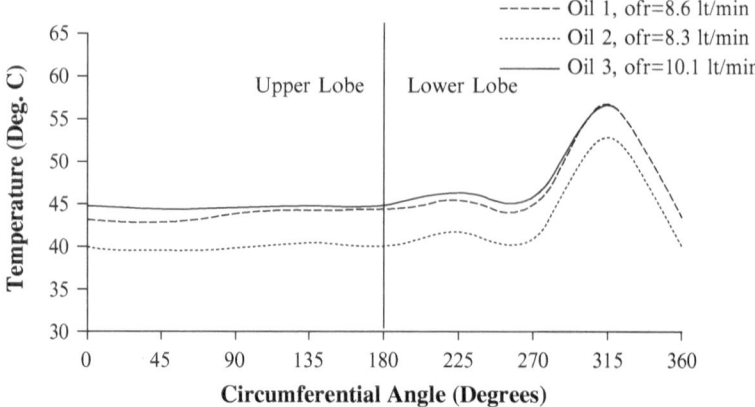

Fig. 3.17 Circumferential variation of temperature (experimentally obtained) in the central plane of offset-halves journal bearing at 3500 rpm and 0.3 MPa pressure at load = 300 N for different grade oils

References

Chauhan A. Experimental and theoretical investigations of the thermal behaviour of some non-circular journal bearing profiles, Ph.D. thesis, Mechanical Engineering Department, NIT Hamirpur (H.P.), India; 2011.

Edgar Jr JG. Rotor bearing geometry. In: Proceedings of the first turbomachinery symposium. Texas A&M University, Oct. 1972. p. 119–41.

Hori Y. Hydrodynamic lubrication. Tokyo: Springer; 2006.

Hussain A, Mistry K, Biswas S, Athre K. Thermal analysis of non-circular bearing. Trans ASME J Tribol. 1996;118:246–54.

Kasolang S, Dwyer-Joyce RS. Observations of film thickness profile and cavitation around a journal bearing circumference. Tribol Trans. 2008;51:231–45.

San Andres L. Hydrodynamic fluid film bearings and their effect on the stability of rotating machinery. In: Design and analysis of high speed pumps. Educational notes RTO-EN-AVT-143, paper 10. Neuilly-sur-Seine, France: RTO; 2006. p. 10-1–10-36.

Sehgal R, Swamy KNS, Athre K, Biswas S. A comparative study of the thermal behaviour of circular and non-circular journal bearings. Lubr Sci. 2000;12:329–44.

Sharma RK, Pandey RK. Effects of the temperature profile approximations across the film thickness in thermohydrodynamic analysis of lubricating films. Indian J Tribol. 2007;2:27–37.

Stachowiak GW, Batchelor AW. Engineering tribology. The Netherlands: Elsevier Science Publishers B.V.; 1993. p. 123–31.

Chapter 4
Review of Literature

This section provides details of research (both theoretical and experimental) carried out on hydrodynamic bearings, both circular and non-circular geometry. Among non-circular profiles, the research was carried out on offset-halves, multi-lobe, and elliptical journal bearings in particular. The literature also includes the different methods involved in the thermal analysis of the journal bearing. Enormous information on the theoretical and experimental work related to the circular journal bearings have been observed in the literature. However, such works pertaining to non-circular journal bearings, especially for offset-halves, multi-lobe, and elliptical journal bearings, are limited, and hence, are the main areas of focus in the present study. It has been a known fact that friction can be reduced with the use of suitable lubricant from the earlier days. However, during the nineteenth century, the mechanism pertaining to lubrication became known to all when the expansion of railroad system in United Kingdom commenced. Tower was assigned to investigate rail axle-bearing problems by the Institution of Mechanical Engineers (UK), and during investigation, he observed that the maximum pressure developed in journal bearing was six times higher than the mean bearing pressure, whereas the peaks were shifted towards the direction of motion. Further, Osborne Reynolds' (1886) findings on hydrodynamic bearings motivated the researchers to go for research in areas of hydrodynamic journal bearings. The pressure generated in the lubricant due to its motion is called hydrodynamic pressure. Many researchers have reported that the oil film temperature rise is higher for circular journal bearings when compared to non-circular journal bearings. It has been also reported by the researchers like Hussain et al. (1996), Ma and Taylor (1996), Sehgal et al. (2000), Chauhan and Sehgal (2008), Chauhan et al. (2010) and few others that the non-circular journal bearings are quite stable and they run cooler during operation. In this section, the theoretical and experimental works pertaining to non-circular journal bearings have been summarized:

© The Author(s) - SpringerBriefs 2016
A. Chauhan, *Non-Circular Journal Bearings*, SpringerBriefs in Materials,
DOI 10.1007/978-3-319-27333-4_4

4.1   Theoretical Studies

Pinkus and Lynn (1956a, b) have derived the power losses for elliptical and three-lobe bearings, both symmetrical and asymmetrical as functions of the bearing parameters and bearing ellipticity. Authors have given expressions for two cases. In the first one by assuming a complete oil film, and in the second one by taking into account the incompleteness of the oil film in the diverging sections of the bearing. They have also presented the analysis of elliptical bearings based on the numerical solution of Reynolds equation for finite bearings. The authors have supplemented the solution of differential equation by focusing on the nature of the oil flow, power loss, and eccentricity in elliptical journal bearings. Wilcock (1961) has worked towards the possibility of displacing the lobe centers of two-lobe journal bearings orthogonally with respect to the mid-radius of the lobe. The author shows that when the lobe displacement is in a direction opposite to the shaft surface motion, and the bearing is centrally loaded, shaft stiffness orthogonal to the load vector is substantially increased. At the same time, vertical stiffness essentially remains unchanged and minimum film thickness is decreased; particularly at low loads, while oil flow is increased. Author also carried out an analysis for a bearing having in cross-section two arcs (each subtending an angle of 150°), $L/D = 1/2$, and with the arc centers each displaced from the geometric center by half the radial clearance. Black and Murray (1974) have described a theory which allows the characteristics of bearings operating in the laminar or turbulent regimes to be evaluated by a similar method, using less storage requirements than finite difference methods and bearings of different geometries can be easily analysed using the program structure. When multi-bearing configurations are being considered, the load magnitudes and directions are dependent on the bearing characteristics and cannot be directly calculated. The authors prepared a databank to provide information (1883) on circular, partial arc, offset halves, and lemon bore bearings operating in the laminar and turbulent regimes, together with a fast interpolation subprogram. Singh et al. (1977) have reported that non-circular bearings are finding extensive use in high speed machinery since they enhance shaft stability, reduce power losses, and increase oil flow characteristics, thus reducing bearing temperature. The authors had presented a solution to analyse the elliptical bearings, using a variational approach.

Sinhasan et al. (1980) have analysed two-lobe porous hydrodynamic journal bearing to study its static and dynamic performance characteristics. A comparison of the static and dynamic performance characteristics of porous and non-porous two-lobe bearings has also been made by the authors on the basis of comprehensive data for a wide range of permeability parameters. The authors observed that the performance of porous bearings is inferior to that of non-porous bearings. Malik et al. (1981) have presented the performance characteristics of tilted three-lobe journal-bearing configurations. This investigation of the performance characteristics over a wide range of tilt angles shows that, within a specific range of tilt angles, the tilted configurations exhibit superior dynamic performance to the usual symmetric three-lobe bearing. Later, Malik et al. (1982) have carried out the task of providing comprehensive design data, which include both static and dynamic performance characteristic for the offset-halves journal bearing. All the design data reported by the authors has been obtained for an offset factor of 0.5 with four aspect ratios (1/3, 1/2, 2/3, and 1). Further, the design data cover laminar as well as turbulent regimes of operations with four values of Reynolds number (3000, 6000, 9000, and 12,000). The authors have also described the analytical procedure for obtaining the performance characteristics of the bearing. Singh and Gupta (1982) have considered the stability limits of elliptical journal bearings supporting flexible rotors. The Reynolds equation has been solved numerically for several values of the eccentricity ratios, L/D ratio, and dimensionless velocity of the journal centre. The authors have observed that the operating load, ellipticity, L/D ratio, and shaft flexibility significantly affect the limits of stable operation. The authors have also reported that elliptical bearings are suitable for stiffness and for moderately flexible rotors. Booker et al. (1982) have concluded that 'offset' designs offer greatly improved durability and reduced power loss in applications for which conventional journal bearings are marginally satisfactory. Authors have reported that these designs are attractive for duty cycles which combine non-reversing loading with limited angular oscillation. The presented analysis combines impedance with a generalized short bearing film model for partial arc bearings. Chandra et al. (1983) have compared the performance of different bearing configurations, namely offset-halves, lemon-bore, three-lobe, and four-lobe bearing at the same load capacity and speed. During the comparison, the authors have considered the steady state and stability characteristics. Booker and Govindachar (1984) have suggested that Novel 'offset' designs offer attractive possibilities in several applications for which conventional journal bearings are only marginally satisfactory. They considered one such problem in rotating machinery (the support of a rigid (massive) rotor turning at high speed under a fixed (gravity) load). The problem has been studied by the authors through a numerical example as well as through parametric studies. The authors show that the stability of full journal bearing system is significantly improved by moderate offset and is fairly insensitive to small departures from optimal design values. Booker and Olikara (1984) in another work have suggested that offset designs are particularly attractive for duty cycles, which combine non-reversing loadings with limited angular oscillations. The authors reported that the offset designs seem equally promising for steady state operations with counter-rotations of shaft and sleeve under fixed load, or with load rotating at half shaft speed. Both cases have been studied by the authors through numerical examples and parametric studies. It has been shown by the authors that the performance of full journal bearings has been significantly improved by small offsets and fairly insensitive to small departures from optimal values. Hashimoto and Wada (1984) have studied the performance characteristics of elliptical journal bearings in turbulent flow regime theoretically as well as experimentally. It has been observed by the authors that the turbulence significantly affects the static and dynamic characteristics of such bearings and the stability limits of such bearings become larger with an increase of ellipticity ratio. Singh and Gupta (1984) have theoretically predicted the stability of a hybrid two-lobe bearing, which is obtained by displacing the lobe centers of a circular profile which leads to non-circular profile named elliptical bearing. It has been found by the authors that an orthogonally displaced bearing is more stable than the known bearings and at the same time it is easier to manufacture. Martin and Ruddy (1984) have reported that profile bore journal bearings are used to suppress rotor instability caused by oil whirl. The effect of manufacturing tolerances on the clearances in a tilted three-lobe bearing and an offset halves bearing is demonstrated by the authors. Results presented by the authors show the effects of these dimensional variations and how these changes affect the instability threshold speed. Mehta and Singh (1986) have analysed the dynamic behaviour of a cylindrical pressure dam bearing in which centers of both halves are displaced. Authors observed that the stabilities of a cylindrical pressure dam bearing can be increased many times by displacing the centers of two halves. It has been reported by the authors that the bearing so obtained is even superior to elliptical and half elliptical pressure dam bearings in stability. Nair et al. (1987) have presented the effects of bearing deformation on the static and dynamic performance characteristics of an elliptical bearing. The finite element method has been used by them to solve the three-dimensional Navier–Stokes, continuity, and elasticity equations. The characteristics presented in this paper take into account the flexibility of the bearing liner.

Basri and Gethin (1990) have carried out a theoretical analysis of the thermal behaviour of orthogonally displaced, three-lobe, and four-lobe bearing geometries. The thermal analysis illustrates the implication of the type selection with regard to the parameters of load-carrying ability, power loss, lubricant requirements, and operating temperatures. The comparisons presented by the authors show that all profiles considered have inferior load-carrying ability and less extreme thermal effects when compared with the cylindrical geometry along with significantly larger lubricant supply requirements. Crosby (1991) has provided a thermohydrodynamic (THD) solution of the two-lobe bearing considering reverse flow at the leading and trailing edge. The author has obtained the temperature and pressure distributions by the simultaneous solutions of the generalized Reynolds equation and the Energy equation, taking into consideration the probable existence of a reverse flow region at the leading edge. Pai and Majumdar (1992) have determined the stability of submerged four-lobe oil journal bearings under dynamic load. The authors used a non-linear transient method to predict the journal centre trajectory for a submerged four-lobe oil journal bearing under (1) unidirectional constant load, (2) unidirectional periodic load, and (3) variable rotating load. They integrated Reynolds equation using the Jakobsson–Floberg–Olsson cavitation zone model. Authors observed that the excursions of the journal centre were subdued, unlike for the plain cylindrical bearings, where the journal centre had large excursion before it became stable or ended in a limit cycle. Interesting trajectories have been observed by the authors for the periodic load. Later on, Crosby (1992) analysed numerically a finite journal bearing having a slightly elliptical bush, subjected to prescribed non-dimensional loads ranging from 0.05 to 0.5. The author has investigated the two effects on the performance of elliptical bearing. First, the ellipticity effect and the second, the angle between the major axis of the bush and the load line on these operating characteristics. It has been reported by the author that the optimum value of inclination angle in order to minimize the maximum pressure and friction torque are 30°–65° and 110°, respectively. Mehta and Rattan (1993) have reported that the use of multi-lobe bearings is considered to be more stable than ordinary circular bearings in industry and its usage is increasing daily and the studies have shown that performance can be further improved if pressure dams are incorporated into these bearings. The authors carried out an analysis to assess the performance of a three-lobe bearing with pressure dams. The results show that the performance of a three-lobe pressure-dam bearing is far superior to that of an ordinary three-lobe bearing. Nagaraju et al. (1994) presented a THD solution for a finite two-lobe journal bearing and obtained temperature and pressure distributions by simultaneous solutions of the generalized Reynolds equation, the Energy equation, and the heat conduction equation. The authors computed static characteristics (in terms of load capacity, attitude angle, end leakage, and friction parameter) and the dynamic characteristics (in terms of critical mass, threshold speed, and damped frequency of whirl). Hussain et al. (1996) have predicted the temperature distribution in various non-circular journal bearings, namely two-lobe, elliptical, and orthogonally displaced. The work reported is based on a two-dimensional treatment following McCallion's approach (1970) in which the Reynolds and energy equations in oil film are decoupled by neglecting all pressure terms in the energy equation. It has been concluded by the authors that the temperature rise is highest for the two-lobe bearing followed by elliptical and orthogonally displaced bearings. Gethin (1996) has presented a numerical investigation into the effect of temperature boundary conditions and viscous dissipation on the behaviour of two- and three-lobe bearings. Model results are compared with experimental data and they introduce the need to include temperature fade at the downstream edge of the lobe. This has been included as a temperature prescription and good agreement has been achieved between numerical and experimental results for three-lobe geometry by them. The authors have reported that, for the two-lobe bearing, by setting viscous generation to zero in the cavitated film region gave the best agreement with experiment. This supports the hypothesis that the lubricant adheres to the journal and is carried by it in the cavitated film. In another study, Vincent et al. (1997) have presented a numerical investigation of cavitation in dynamically loaded journal bearings using mobility method. The authors have considered the mobility method of Booker to improve the computing efficiency. The authors have presented results for crankshaft bearing of the Ruston and Hornsby diesel engine and have considered the effect of non-circular profile (elliptical) on film rupture and reformation during the analysis. Ma and Taylor (1999) have presented a theoretical evaluation of two-lobe circular, elliptical, offset-half, three-lobe, and four-lobe bearings through a comparison of their steady state performance characteristics such as minimum film thickness, maximum temperature, power loss, and flow rate using THD model. The results presented by the authors show that the performance of the non-circular bearings is inferior to that of circular bearing; of non-circular types, the offset-half and elliptical bearings have better overall performances.

Pereira et al. (2000) have analysed the flow between rotating eccentric cylinders with axial throughput. The authors have modeled the axial throughput by flow driven by a constant pressure gradient. The authors have also studied the effects of inertia at values typical for the device and conditions for stress-induced cavitation have been evaluated. The flow is completely determined by a Reynolds number, an eccentricity ratio, and a dimensionless pressure gradient and all computed results are either presented or can be easily expressed in terms of these dimensionless parameters. It has been reported that the effect of inertia is to shift the eddy or recirculation zone which develops in the more open region of the gap toward the region of low relative pressure, zero of the relative pressure migrates away from the center, and the distribution breaks the skew symmetry of the Stokes flow solution. Strzelecki (2000) has presented data on the maximum oil film pressure and temperature of two-lobe journal bearings with different bore profiles. A bearing with a cylindrical non-continuous bore profile and another with a pericycloid continuous profile have been evaluated by the author. Authors have determined the minimum values of oil film thickness, maximum pressure values, and temperatures of the oil film for an assumed relative length of the bearings, and parallel orientation of the journal and bearing axes. The calculations carried out are based on the assumption of an adiabatic oil film and a static equilibrium position of the journal. Sehgal et al. (2000) have presented a comparative theoretical analysis of three types of hydrodynamic journal-bearing configurations, namely, circular, axial groove, and offset-halves. It has been observed by the authors that the offset journal bearing runs cooler than an equivalent circular journal bearing with axial grooves. Hashimoto and Matsumoto (2001) have described the optimum design methodology for improving operating characteristics of hydrodynamic journal bearings and its application to elliptical journal-bearing design used in high-speed rotating machinery. The hybrid optimization technique combining the direct search method and the successive quadratic programming has been applied effectively by the authors to find the optimum solutions. In the optimum design of elliptical journal bearings, authors have determined the design variables such as vertical and horizontal radial clearances, bearing length to diameter ratio, and bearing orientation angle to minimize the objective function defined by the weighted sum of maximum averaged oil film temperature rise, leakage flow rate, and the inversion of whirl onset speed of the journal under many constraints. Comparing the optimized operating characteristics with the characteristics calculated from the randomly selected design variables, the effectiveness of optimum design is clarified by the authors. Strzelecki (2001) has presented the calculations for load capacity of two-lobe journal bearings characterized by different profiles of upper and bottom lobe. In the analysis, author has assumed adiabatic oil flow in the bearing lubricating gap, parallel orientation of journal and bearing axis as well as the static equilibrium position of the journal. The author has concluded that the two-lobe type (upper half offset and bottom one cylindrical) bearing has the largest value of load carrying capacity among the all considered bearing profiles. Mishra et al. (2007) have considered the non-circularity in bearing bore to be elliptical and the same is compared with circular bearing to analyse the effect of irregularity. Authors have presented the numerical solution of Reynolds equation and Energy equation for an elliptic bore journal bearing to outline the temperature profile and have solved the Energy equation adiabatically. It has been reported by the authors that with increasing non-circularity, the pressure gets reduced and temperature rise is less in case of journal bearing with higher non-circularity value. Mishra (2007) has analysed an irregular bore journal bearing of elliptic shape for various performance parameters considering the adiabatic temperature rise in the lubricant conjunction. The author has investigated the ratio of thermal to isothermal value of all parameters for a range of non-circularity value with various eccentricity ratios. The author has also plotted the temperature contour to study the behaviour of isotherms due to the presence of bore ellipticity. Wang and Damodaran (2007) have solved two-dimensional incompressible Navier–Stokes equations numerically to model the performance of a dynamically loaded journal bearing and the bearing under study has been modeled as eccentrically rotating cylinders. The authors have introduced a single domain pseudospectral method which combines Fourier expansions and Chebyshev polynomials for spatial discretization in conjunction with appropriate time marching scheme for the unsteady incompressible Navier–Stokes equations. The pseudo-spectral scheme is applied by the authors to a few classical problems: concentric rotating cylinders and journal bearings with lubricants of constant and varying viscosity to establish the validity of the numerical scheme in simulating these problems realistically as well as to gauge the convergence characteristics and relevant numerical issues. The numerical modeling proposed by the authors has been found to be reasonably accurate and robust enough to serve as a tool for the study of flow in the region between the journal and bearing. Deng and Braun (2008) have investigated the flow of a viscous fluid in the narrow gap ($C/R = 0.002$–0.01) between two cylinders with low eccentricity ratios ($\epsilon \leq 0.2$), where the shaft radius is 25.4 mm (1.0 in.). The computational engine is provided by CFD-ACE+, commercial multi-task software. The analysis presented by the authors for the Navier–Stokes equation shows that the inertia, viscous terms, pressure, and the Reynolds stress terms are equally significant. Based on these findings, they have proposed a new model for predicting the flow behaviour in long journal-bearing films in the transition regime (Taylor and wavy vortex regimes) and justified. The new model indicates that the velocity profiles are sinusoidal and depend on the local Reynolds number and the position in the axial direction. Finally, a modified form for the Reynolds equation has been proposed by the author for the transition regime. Gertzos et al. (2008) have reported that there is an increasing interest in designing hydrodynamically lubricated bearings using electro-rheological fluids (ERFs) or magnetorheological fluids (MRFs). Both smart fluids behave like Bingham fluids, and thus the Bingham plastic model is used to describe the grease and the ERFs and MRFs behaviour of the non-Newtonian fluid flow. The performance characteristics of a hydrodynamic journal bearing lubricated with a Bingham fluid are derived by means of three-dimensional computational fluid dynamics analysis. Journal-bearing performance characteristics, such as relative eccentricity, attitude angle, pressure distribution, friction coefficient, lubricant flow rate, and the angle of maximum pressure, are derived and presented for several length over diameter (L/D) bearing ratios and dimensionless shear numbers of the Bingham fluid. The diagrams are presented in the form of Raimondi and Boyd charts and can easily be used in the design and analysis of journal bearings lubricated with Bingham fluids. The authors have developed design charts that could be used by the designer engineer to design smart journal bearings. Crosby and Chetti (2009) have studied the static and dynamic characteristics of two-lobe journal bearings lubricated with couple-stress fluids. The authors have observed that the effect of the couple stress parameter is significant on the performance of the journal bearing and also the stability gets improved compared to the bearings lubricated with Newtonian fluids. Ostayen and Beek (2009) have carried out a finite element analysis of a lemon-bore journal bearing. The THD model presented by the authors is an inverse model, that is, a model in which the shaft eccentricity and attitude angle are calculated by giving a certain known and prescribed load and load angle. In an analysis carried out by authors, care has been taken to accurately model the flow of heat to and from the oil supplies and the model is used to check the design of the lemon bearings in a specific naval application.

Liu et al. (2010) have simulated three different journal bearing models: a pure fluid bearing model, a fluid structure interaction (FSI) squeeze-film model, and a bearing-rotor FSI model using a CFD and FSI technique and investigated interaction of the lubrication of the journal bearing and the dynamics of the shaft. The combined CFD and FSI method has been subsequently used by the authors to study the lubrication performance of the rotor-bearing system. An elastic shaft was used in the full coupled bearing-rotor FSI model, as well as the Gümbel boundary condition. The loads applied on the model are: vertical load and a rotation which is representative of real working conditions of an experiment of marine journal bearing. The authors have observed a good agreement with the already published results and suggested that more complex models will be developed to investigate more realistic rheological properties of the lubricant and the complex interactions using CFD and FSI methods. Rahmatabadi et al. (2010) have considered the non-circular bearing configurations: two-, three- and four-lobe lubricated with micropolar fluids. The authors have modified the Reynolds equation based on the theory of micropolar fluids and solved the same by using finite element methods. It has been observed by the authors that micropolar lubricants can produce significant enhancement in the static performance characteristics and the effects are more pronounced at larger coupling numbers. Xing and Braun (2012) have reported that to determine the stiffness, damping, and added mass coefficients of the hydrodynamic bearing, the finite perturbation method around its stabilization position was employed. Authors have used the full three-dimensional Navier–Stokes equations in CFD-ACE+ to evaluate the dynamic coefficients from an actual lubricant and compared to those obtained with Reynolds equation. Finally, a homogeneous gaseous cavitation algorithm is coupled with the Navier–Stokes equation to establish the pressure distribution in the bearing. It has been found that when gas concentration varied the pressure distribution and the dynamic coefficients changed significantly, it can significantly influence the performance of the bearing. Since it's difficult to study the fluid/structure interactions in a rotor-bearing system using the conventional method, a new transient analysis method combining computational fluid dynamics and fluid–structure interaction was applied based on actual physical model by Lin et al. (2013). A comparison with the published experimental results has been presented and discussed, and theoretical predictions found to be agreed well with the experimental ones. The authors have suggested that the developed method can be a very useful tool for the study on the bearing lubrication problem and can effectively and accurately predict the transient lubrication process. Considering the differences between the physical properties of the water and of the oil, the effects of eccentricity ratio on pressure distribution of water film are analysed by computational fluid dynamics (Gao et al. (2014)). Based on the analysis, a reference is produced by the authors for selecting the initial diameter dimension which is used to design an efficient water-lubricated plain bearing under the given conditions of load and rotational speed. Also the reference has been verified by an experimental case by the authors. The performance of circular, two-lobe, and elliptical journal bearing by TEHD and THD analysis has been presented by Huang et al. (2014). It has been found that under the identical geometric parameters and operating condition, the circular journal bearing possesses the greatest magnitude of the maximum oil film pressure, the two-lobe one takes the second place, and the elliptical one possesses the smallest magnitude. The thermo-elasto deformations on the pad are the same order of magnitude with the minimum film thickness. Vakilian et al. (2014) have investigated the THD characteristics of Rayleigh step bearings running under steady, incompressible, and laminar conditions. An attempt has also been made to investigate the THD behaviours of step bearings when they are running under various operating conditions like different speeds, geometrical factors, and Reynolds numbers. Discretized forms of mass, momentum, and energy equations are obtained by the finite volume method and solved using the SIMPLE algorithm. In addition, an attempt has been made to investigate the effects of runner surface velocity and bearing geometrical factors on the lubricant pressure, temperature distributions, load capacity, and friction force. Authors have observed a good consistency of numerical results with theoretical findings in the open literature.

Aksov and Aksit (2015) have developed a model to predict three-dimensional thermal, structural, and hydrodynamic performance of foil bearings. Bearing deformation and film pressure have been coupled in FEM by the authors considering thermal/centrifugal growths and thermo-mechanical material properties. Authors have applied Augmented-Lagrangian method and thermal contact models to solve for mechanical and thermal contacts. A parametric study has been presented by the authors for various shaft speeds and load conditions to visualize the bearing performance. A numerical study of the hydrodynamic lubrication between two parallel surfaces with micro-texturing has been presented by Caramia et al. (2015). The two-dimensional Navier–Stokes equations for an isothermal incompressible steady flow have been considered as a suitable model. A detailed analysis of flow velocity profiles and pressure distributions has been presented to study the forces acting on the textured surface. Furthermore, results show that depending on the cavity depth, three regions exist with a different flow dynamics. Finally, the authors have found an optimal value of the depth for which the pressure reaches a minimum value and the probability of cavitation is maximized. Gengyuan et al. (2015) have reported that water-lubricated bearing is becoming more and more popular since it is environmentally friendly and saves energy. However, contrary to oil- and grease-lubricated bearings, water-lubricated bearing is limited in many situations due to its low hydro-dynamic load-carrying capacity. The authors have proposed a new bearing bush, with a transition-arc structure, which is favorable for increasing hydrodynamic load-carrying capacity. Hydrodynamic load-carrying capacity was calculated by means of three-dimensional computational fluid dynamics (3-D CFD) analysis. From the results, authors have observed the changes in hydrodynamic load-carrying capacity of a water-lubricated journal bearing. For different width over diameter (L/D)-bearing ratios, the relationship between hydrodynamic load-carrying capacity and the magnitude of the transition-arc structure dimension has also researched.

4.2   Experimental Studies

Flack et al. (1980) have investigated the pressures in four-lobe bearings, both exper-imentally and theoretically. The authors tested a four-lobed bearing 25.4 mm in diameter with the load vector 'on pad' and 'off pad'. Static pressures were measured on the centre line of the bearing and these experimental data are compared with two sets of theoretical results. The authors used Half-Sommerfeld and Reynolds bound-ary conditions in the theoretical predictions. It has been observed by the authors that the trends of the pressure versus rotational speed for the experimental data and the theoretical solution are the same for the Half-Sommerfeld condition, but sometimes differ for the Reynolds condition. Read and Flack (1987) have developed a test apparatus on which an offset-halves journal bearing of 70 mm diameter journal was tested at five vertical loads and two rotational speeds. The authors have determined the pressure, temperature, and film thickness profiles around the bearing. The authors have observed that the data follow the expected trends. Fitzgerald and Neal (1992) have presented some experimental data for 76 mm diameter two-axial groove circular bearings. According to their results, the axial temperature variation was negligible, but the circumferential temperature variations are very significant. Basri and Gethin (1993) have completed some experimental investigation into the thermal behaviour of a three-lobe profile bore bearing. The authors considered three geom-etries, namely symmetric, tilted, and tilted with side rail. Global behaviour is also presented by the authors, which demonstrates the trends similar to those with a cylindrical geometry. The results confirm the effectiveness of the side rail in reduc-ing lubricant side flow while having only a marginal effect on thermal excursions. Ma and Taylor (1996) have experimentally investigated the thermal behaviour of a two-axial-groove circular bearing and an elliptical bearing, both 110 mm in diam-eter. Both bearings were tested by the authors at specific loads upto 4 MPa and rotational frequencies up to 120 Hz. The authors have measured power losses and flow rates directly and detailed temperature information was collected. The result presented by the authors show that the thermal effects are significant in both bear-ings. Makino et al. (1996) have described the performance characteristics of two types of journal bearings: tilting-pad and offset-halves, operating under high-speed conditions where the journal peripheral speed exceeds 100 m/s. Experimental investigations into bearing eccentricity, temperatures, and dynamic coefficients have been carried out with a specially designed test rig. These results are compared with the theoretical data which are obtained from THD lubrication theory and are found satisfactory. Arumugam et al. (1997) have experimentally evaluated the performance characteristics of a misaligned three-lobe journal bearing. The parameters which have been studied include friction, vibration responses, minimum film thickness, stiffness, damping coefficients of the fluid film, system natural frequency, and damping factor. Authors observed that the film thickness decreases, friction increases, and system damping increases as the bearing misalignment increases.

4.3   Research Gaps

Based on the above literature survey, the following conclusions have been drawn:

- Experimental studies of oil film temperature across the width of offset-halves, multi-lobe, and elliptical journal bearing are reported to very small extent in literature.
- Thermal effects in non-circular bearings, especially in Offset-halves, Orthogonally Displaced, Lemon bore, Two-lobe, Three-lobe, Multi-lobe, and Elliptical journal-bearing configurations, have been explored only to a very small extent. However, these bearings find wide applications in high speed machinery with light loads because of their better stability and cool running characteristics.
- Literature reveals that formulation of Energy equation for non-circular journal bearings with parabolic temperature profile technique is done for some of the bearings only.
- Recently, few authors are using simulation software for thermal study of circular journal bearings and very limited have used these software for non-circular journal bearings. Author of the current paper finds it difficult to find any paper related to oil film temperature for the bearings under study using such simulation softwares.

References

Aksov S, Aksit MF. A fully coupled 3D thermo-elastohydrodynamics model for a bump-type compliant foil journal bearing. Tribol Int. 2015;82:110–22.

Arumugam P, Swarnamani S, Prabhu BS. Effects of journal misalignment on the performance characteristics of three-lobe bearings. Wear. 1997;206:122–9.

Basri S, Gethin DT. A comparative study of the thermal behaviour of profile bore bearings. Tribol Int. 1990;23:265–76.

Basri S, Gethin DT. An experimental investigation into thermal behaviour of a three-lobe profile bore bearings. J Tribol-T ASME. 1993;115:152–9.

Black HF, Murray JL. Calculation and selection of dynamic properties of journal bearings suitable for high speed applications. IEE Conference Publication (London) England, April 8–11, 1974;49–56.

Booker JF, Govindachar S. Stability of offset journal bearing systems. Proc Inst Mech Eng. 1984;C283/84:269–75.

Booker JF, Olikara P. Dynamics of offset bearings: parametric studies. J Tribol-T ASME. 1984;106:352–9.

Booker JF, Goenka PK, Van Leeuwen HJ. Dynamic analysis of rocking journal bearing with multiple offset segments. J Lubric Tech-T ASME. 1982;104:478–90.

Caramia G, Carbone G, Palma PD. Hydrodynamic lubrication of micro-textured surfaces: two dimensional CFD-analysis. Tribol Int. 2015;88:162–9.

Chandra M, Malik M, Sinhasan R. Comparative study of four gas-lubricated non-circular journal bearing configurations. Tribol Int. 1983;16:103–8.

Chauhan A, Sehgal R. An experimentation investigation of the variation of oil temperatures in offset-halves journal bearing profile using different oils. Indian J Tribol. 2008;2:27–41.

Chauhan A, Sehgal R, Sharma RK. Thermohydrodynamic analysis of elliptical journal bearing with different grade oils. Tribol Int. 2010;43:1970–7.

Crosby WA. A thermohydrodynamic solution of the two lobes bearing considering reverse flow at the leading and trailing edges. Wear. 1991;143:159–73.

Crosby WA. An investigation of the performance of a journal bearing with a slightly irregular bore. Tribol Int. 1992;25:199–204.

Crosby WA, Chetti B. The static and dynamic characteristics of a two-lobe journal bearing lubricated with couple-stress fluid. Tribol Trans. 2009;52:262–8.

Deng D, Braun MJ. A new model for transition flow of thin films in long journal bearings. Tribol Trans. 2008;51:1–11.

Fitzgerald MK, Neal PB. Temperature distributions and heat transfer in journal bearings. J Tribol-T ASME. 1992;114:122–30.

Flack RD, Leader ME, Allaire PE. An experimentally and theoretical investigation of pressures in four-lobe bearings. Wear. 1980;61:233–42.

Gao G, Yin Z, Jiang D, Jiao XZS. Numerical analysis of plain journal bearing under hydrodynamic lubrication by water. Tribol Int. 2014;75:31–8.

Gengyuan G, Zhongwi Y, Dan J, Xiuli Z. CFD analysis of load-carrying capacity of hydrodynamic lubrication on a water-lubricated journal bearing. Ind Lubr Tribol. 2015;67:30–7.

Gertzos KP, Nikolakopoulos PG, Papadopoulos CA. CFD analysis of journal bearing hydrodynamic lubrication by Bingham lubricant. Tribol Int. 2008;41:1190–204.

Gethin DT. Modelling the thermohydrodynamic behaviour of high speed journal bearings. Tribol Int. 1996;29:579–596.

Hashimoto H, Matsumoto K. Improvement of operating characteristics of high-speed hydrodynamic journal bearings by optimum design: part I-formulation of methodology and its application to elliptical bearing design. J Tribol-T ASME. 2001;123:305–12.

Hashimoto H, Wada S. Performance characteristics of elliptical journal bearings. B JSME. 1984;27:2265–71.

Huang B, Wang L, Guo J. Performance comparison of circular, two-lobe and elliptical journal bearings based on TEHD analysis. Ind Lubr Tribol. 2014;66:184–93.

Hussain A, Mistry K, Biswas S, Athre K. Thermal analysis of non-circular bearing. J Tribol-T ASME. 1996;118:246–54.

Lin Q, Wei Z, Wang N, Chen W. Analysis on the lubrication performances of journal bearing system using computational fluid dynamics and fluid–structure interaction considering thermal influence and cavitation. Tribol Int. 2013;64:8–15.

Liu H, Zhongmin J, Xu H, Ellison P. Lubrication analysis of journal bearing and rotor system using CFD and FSI techniques. In: Jianbin L, Yonggang M, Tianmin S, Qian Z, editors. Advanced tribology. Berlin: Springer; 2010. p. 40–1.

Ma MT, Taylor CM. An experimental investigation of thermal effects in circular and elliptical plain journal bearings. Tribol Int. 1996;29:19–26.

Ma MT, Taylor CM. A comparative thermal analysis of the static performance of different fixed profile bore plain bearings. Proc Inst Mech Eng J-J Eng. 1999;213:13–30.

Makino T, Morohoshi S, Taniguchi S. Thermohydrodynamic performance of high-speed journal bearings. Proc Inst Mech Eng J-J Eng. 1996;210:179–88.

Malik M, Chandra M, Sinhasan R. Performance characteristics of tilted three-lobe journal bearing configurations. Tribol Int. 1981;14:345–9.

Malik M, Chandra M, Sinhasan R. Design data for offset-halves journal bearings in laminar and turbulent regimes. ASLE Trans. 1982;25:133–40.

Martin FA, Ruddy AV. Effect of manufacturing tolerances on the stability of profile bore bearings. Third international conference on vibrations in rotating machinery, Heslington, England; 1984. p. 287–93.

McCallion H, Yousif F, Lloyd T. The analysis of thermal effects in a full journal bearing. Transactions of the ASME, J Lubr Technol. 1970;92:578–587.

Mehta NP, Rattan SS. Performance of three-lobe pressure-dam bearings. Tribol Int. 1993;26:435–42.

Mehta NP, Singh A. Stability analysis of finite offset-halves pressure dam bearing. J Tribol-T ASME. 1986;108:270–4.

Mishra PC. Thermal analysis of elliptic bore journal bearing. Tribol Trans. 2007;50:137–43.

Mishra PC, Pandey RK, Athre K. Temperature profile of an elliptic bore journal bearing. Tribol Int. 2007;40:453–8.

Nagaraju Y, Joy ML, Nair KP. Thermohydrodynamic analysis of a two-lobe journal bearing. Int J Mech Sci. 1994;36:209–17.

Nair KP, Sinhasan R, Singh DV. Elastohydrodynamic effects in elliptical bearing. Wear. 1987;118:129–45.

Ostayen RAJV, Beek AV. Thermal modelling of the lemon-bore hydrodynamic bearing. Tribol Int. 2009;42:23–32.

Pai R, Majumdar BC. Stability of submerged four-lobe oil journal bearing under dynamic load. Wear. 1992;154:95–108.

Pereira A, McGrath G, Joseph DD. Flow and stress induced cavitation in a journal bearing with axial throughput. J Tribol. 2000;123:1–5.

Pinkus O, Lynn M. Power loss in elliptical and 3-lobe bearings. Transactions of the ASME, Paper No. 54-LUB-9; 1956a. p. 965–73.

Pinkus O, Lynn M. Analysis of elliptical bearings. Transactions of the ASME, Paper No. 55-LUB-22; 1956b. p. 965–73.

Rahmatabadi AD, Nekoeimehr M, Rashidi R. Micropolar lubricant effect on the performance of non-circular lobed bearings. Tribol Int. 2010;43:404–13.

Read LJ, Flack RD. Temperature, pressure and film thickness measurements for an offset half bearing. Wear. 1987;117:197–210.

Reynolds O. On the theory of lubrication and its application to Mr. Beauchamp Tower's experiments, including an experimental determination of the viscosity of olive oil. Philos Trans R Soc London. 1886;177:157–234.

Sehgal R, Swamy KNS, Athre K, Biswas S. A comparative study of the thermal behaviour of circular and non-circular journal bearings. Lubr Sci. 2000;12:329–44.

Singh A, Gupta BK. Stability limits of elliptical journal bearings supporting flexible rotors. Wear. 1982;77:159–70.

Singh A, Gupta BK. Stability analysis of orthogonally displaced bearings. Wear. 1984;97:83–92.

Singh DV, Sinhasan R, Kumar A. A variational solution of two lobe bearings. Mech Mach Theory. 1977;12:323–30.

Sinhasan R, Malik M, Chandra M. Analysis of two-lobe porous hydrodynamic journal bearings. Wear. 1980;64:339–53.

Strzelecki S. Maximum oil film pressure and temperature of two-lobe journal bearings with different bush profiles. Lubr Sci. 2000;12:253–64.

Strzelecki S. Effect of lobe profile on the load capacity of two-lobe journal bearing. Sci China Ser A. 2001;44:95–100.

Tower B. First report on friction experiments. Proc Inst Mech Eng. 1883;1–196:632–59.

Vakilian M, Nassab SAG, Zahra K. CFD-based thermohydrodynamic analysis of Rayleigh step bearings considering an inertia effect. Tribol Trans. 2014;57:123–33.

Vincent B, Maspeyrot P, Frene J. Cavitation in non-circular journal bearings. Wear. 1997;207:122–7.

Wang C, Damodaran M. Numerical modeling of the performance of lubricated journal bearings using Navier-Stokes equations. Int J Comput Fluid D. 2007;14(1):75–96.

Wilcock DF. Orthogonally displaced bearings-I. ASLE Trans. 1961;4:117–23.

Xing C, Braun MJ. Determination of dynamic coefficients in a hydrodynamic journal bearing based on the 3-D Navier-Stokes equations and considering cavitation effects. ASME/STLE 2012 international joint tribology conference, Denver, CO; 2012. p. 7–10.

Chapter 5
Future Work

The chapters have presented the introduction of non-circular journal bearings and how they be analyzed along with the various work that has been carried out to explore the possible use of such bearings in practical application. The possible use of such bearing profiles may be possible once the design charts for these bearings will be developed, and in order to achieve that, the work may be moved in one of the directions listed below:

1. Experiments should be conducted to study the effect of variation of oil inlet pressure on the oil film temperature distribution.
2. Experimentation on various profiles must be conducted to explore their behaviour and the effect of roughness may be introduced while designing the bearing.
3. The aspect of cavitation can also be introduced in the analysis
4. Mathematical modeling using generalized Reynolds equation can be attempted.
5. The work can also be extended by considering the operation of the bearings in turbulent flow region.
6. The CFD-based analysis must be carried out and compared with mathematical modeling to achieve more reliable data.
7. An attempt should be made to develop design charts for non-circular journal bearing.

© The Author(s) - SpringerBriefs 2016
A. Chauhan, *Non-Circular Journal Bearings*, SpringerBriefs in Materials,
DOI 10.1007/978-3-319-27333-4_5